The Biotechnology HANDBOOK for Engineers

Brendan Cooper

Copyright © 2017 Brendan Cooper

All rights reserved.

ISBN-13: 978-1545187463
ISBN-10: 1545187460

CONTENTS

Acknowledgments

1 Facilities

2 Clean Utilities

3 Sterile Manufacturing Operations

4 Depyrogenation

5 Cleaning and Disinfection

6 Process Development

7 Physical Processes

8 Equipment Validation

9 Performance Qualification

10 GMP Basics

11 Data Integrity

ACKNOWLEDGMENTS

To all my professional colleagues throughout the UK and abroad who have encouraged this work and provided valuable input along the way.

Published by Life Science Learning, © Copyright 2017, First Edition

It is the responsibility of individuals, companies and organisations to implement the necessary legal and regulatory requirements relevant to their industry. The author will take no responsibility for the application, accuracy or interpretation of this document.

Contents

Chapter 1: Facilities
Introduction 1
Risk and Impact Assessment 1
Qualification Levels 2
Contamination Control 4
Material Flow 4
Material Transfer 5
Disinfection and Cleaning Agents 5
Gown-Up Areas 5
GMP Zoning 6
Environmental Grade A (Aseptic) 8
Environmental Grade B 8
Environmental Grade C 8
Environmental Grade D 9
Compliance Tests for GMP Zones 9
Environmental Monitoring 10

Chapter 2: Clean Utilities
Introduction 12
Key Definitions 12
Identifying Critical Utilities 12
Compressed Air 13
Water Systems 14
Water for Injection 14
CIP/SIP 14
Clean Steam (Pure Steam) 15
HVAC 16

Chapter 3: Sterile Manufacturing Operations
Introduction 18
Sterility 18
Unit Operations 18
Bioreactor Engineering 19
Downstream Processing 20
Removal Operations 21
Filtration 21
Micro Filtration 21
Centrifugation 22
Precipitation 22

Precipitation Methods 22
Membrane Filtration 23
Raw materials 23
Upstream Processing 24
Filling Operations 24
Container Closure Integrity 26
Isolator Barrier Systems 28
Decontamination Agents 31
Containment 32
Steam Sterilisers 32

Chapter 4: Depyrogenation
What is Depyrogenation? 35
Pyrogens 35
Bacterial Toxins 35
Biological Indicators for Dry Heat 37
Control of Materials 38
Start-up Conditions 39
In-Process Controls 39
Cooling 40
Failure of Depyrogenation 40

Chapter 5: Cleaning and Disinfection
Introduction 41
What Do We Mean by the Term "Cleaning"? 41
Verification Versus Validation 41
Definitions 42
Validation Strategies 44
How Are Acceptance Levels Defined? 45
MACO Calculations 47
Cleaning Procedures and Methods 51

Chapter 6: Process Development
Vial Washers 53
Depyrogenation Tunnels 53
Isolators 54

Chapter 7: Physical Processes
Fluid Flow 59
Classification of Fluids 59
Mixing 59
Vessel Geometry 60

Heat Transfer 60

Chapter 8: Equipment Validation
Introduction 62
Materials of Construction 62
Suggested IQ/OQ Verifications/Tests 63
Depyrogenation Tunnels (Equipment Validation) 65
Isolators (Equipment Validation) 65
Steam Sterilisers (Equipment Qualification) 66
HEPA Filters 66

Chapter 9: Performance Qualification
Depyrogenation 68
Isolators 71
Steam Sterilisers 75
Thermal Validation Systems 76
Re-Qualification 78
Culture Media 78

Chapter 10: GMP Basics
What Is GMP? 79
History of GMP 80
Consequences of GMP Failures 93
Rules of GMP 97

Chapter 11: Data Integrity
Introduction 92
Key Terms 93
The Lifecycle of Data 95
System Categorisation 102
Risk Assessments 104

Glossary

Brendan Cooper

CHAPTER 1

FACILITIES

Introduction

Facilities and utilities qualifications are typically prerequisites to the validation of manufacturing equipment and systems. Much of the activity that deals with establishing a facility or building that is *fit for purpose* is managed under the broad heading of commissioning and qualification (C&Q). The terms C&Q are often used interchangeably and in practice some overlap in activity is expected.

Commissioning can be defined as the planned, documented, and managed engineering approach to the start-up and handover of facilities, systems and equipment to the end-user. It must deliver a safe and functional environment that meets the predefined design and user requirements.

In strict terms, qualification is more concerned with the confirmation and documentation showing that equipment or systems are properly installed and functional. Qualification forms part of validation, but the individual qualification steps do not equal a validated process. The establishment of a user requirements specification (URS) and detailed design specifications ensure that the building or facility will meet end-users' needs and that it is fit for the intended purpose.

It also provides a level of protection to the contracting company responsible for the project or facility construction. Post-URS approval requires an approved Design Qualification (DQ). This provides verification and a documented record that the proposed design is suitable for the intended purpose. Further verification including IQ/OP/PQ should be applied as required based on the system impact and criticality of facilities/utilities.

Risk and Impact Assessment

A risk-based qualification process should assess the potential of a system to impact the product quality. The boundaries of any system (HVAC, compressed air supply etc.) should be identified in order to help establish the scope of any system and determine if it has a direct, indirect or no impact on product quality.

Direct Impact: a system that can directly impact product quality.

Indirect Impact: where a system is not expected to directly impact the product quality but supports or is ancillary to a direct impact system.

No Impact: a system that does not directly impact product quality and does not support a direct impact system.

Example of HVAC System Boundaries

For a HVAC System supplying a classified area, only once the air enters the room must the air quality meet the classified designation. The Critical Quality Attributes (CQAs) are routinely monitored through the Environmental Monitoring Programme and the Critical Process Parameters (CPPs) should be monitored through the calibrated and validated Environmental Monitoring System (QBMS). The direct impact (level 1) for the HVAC systems are indicated on the boundary diagram shown below.

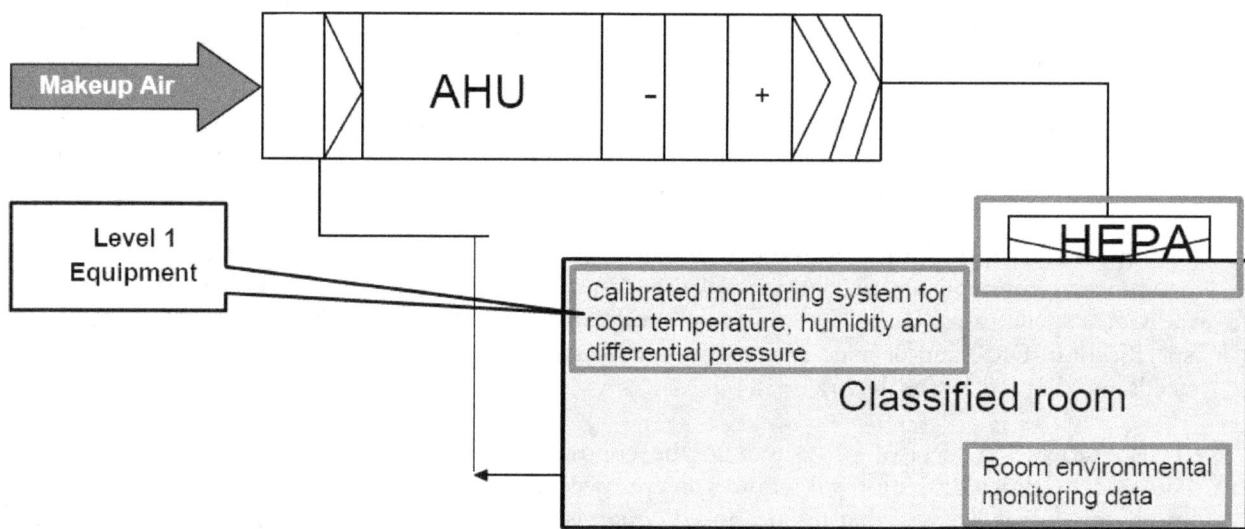

Figure 1: HVAC System Boundary Diagram (Level 1). Each individual system is represented by a green box. Separate qualifications should be performed for each one. The room environmental monitoring system is typically called a Building Management System (BMS). The ***calibrated*** monitoring system for room temperature, humidity and differential pressure is called a QBMS (where the "Q" stands for quality indicating the system is used to monitor critical parameters).

Qualification Levels

Qualification levels are often used within companies to classify the criticality of equipment or systems. Level 1 requires the highest level of verification.

Level 1: a system where an **undetected change** in system performance poses a significant risk to the product and product safety. Level 1 systems require the highest degree of qualification and validation. This should include URS/DQ/IQ/ OQ/ PQ.

Level 2: a system where a change that **may be detected** in system performance poses a significant risk to product and product safety. These systems require a level of qualification including IQ, however, OQ and PQ testing may not be required. This should be based on the intended use of the system, impact on product quality and overall risk.

Level 3: All other systems.

Typically IQ or equivalent testing is sufficient. **Note:** other requirements or qualifications should be based on risk.

The level of qualification and validation testing required for any system should be based on a risk assessment, examining the criticality of the system and environment. Risk assessments should consider the following points:

- Building design and construction features
- System boundaries and complexity
- Potential product impact
- Environmental controls and monitoring systems
- Potential impact to operator safety
- Type of qualification/validation (e.g. prospective, concurrent, or retrospective)

Controlled-not-Classified (CNC) environments, utilities, and facility control systems also require adequate qualification/validation. Again, the impact on product quality should be determined in order to shape any validation. Routine monitoring test locations as well as alert and action levels should be determined in advance of any validation for environmental monitoring or utility systems.

Contamination Control

The philosophy of containment control requires it to be applied across all inputs that make up a facility, equipment, processes, utilities and so on. Containment is primarily concerned with keeping things in by preventing product or processing agents from egressing into the surrounding atmosphere. Ensuring adequate containment protects personnel who interact with the process, equipment and systems. Aseptic processing often deals with biological agents or compounds that may be harmful to operators or technicians. A secondary concern of containment is protection of the environment. Containment also complements efforts in contamination prevention. As with aseptic processing the risk to the patient and product must be at the forefront of activity. Risk-based approaches and tools should be used to identify potential risks and put in place adequate controls and mitigations. Any assessment should take into account all the following systems:

- Facility layout
- Drainage systems
- HVAC requirements
- Location and adequacy of utilities
- Personnel flow and procedures for entering and leaving
- Behavioural requirements of personnel in the clean room
- Flow of materials and products to prevent cross-contamination and mix-ups between products and between dirty and clean or sterile and non-sterile equipment and products
- Design to avoid cross-contamination when manufacturing live biological agents, e.g. local exhaust air HEPA filtration, dedicated air handling units.

Material Flow

The design and layout of any manufacturing area should facilitate the effective flow of materials. This is a fundamental requirement no matter what the industry, e.g. medical devices, pharmaceuticals, bio pharmaceuticals and even non-regulated engineering companies that assemble, machine or fabricate products. However, the manufacture of medicinal products that are required to be sterile imposes a greater level of control and thought. With regard to aseptic processing facilities, material flows do not only require efficient and effective flow of materials; the activity should also support the requirements of aseptic processing while minimising any risk of contamination. Identifying critical processing zones is a crucial step in ensuring the right building design and controls are implemented. Isolators and aseptic filling require the highest classification with strict environmental controls. Secondary packaging operations such as cartonning are often completed in areas controlled and operated to a lower classification.

Design and layout of facilities should:

- Maintain microbiological integrity of the identified critical processing zones
- Prevent or minimise contamination from outside critical processing zones
- Control the flow of materials by restricting access to trained and authorized personnel

Material Transfer

Material transfer from the outside of clean rooms to the inside is completed via material air locks or hatches. Material air locks and hatches ensure that there is clear separation between controlled clean areas and less clean areas. Many suppliers provide products that are double bagged. This provides an added level of control when transferring materials. The outer bag can be removed within the air lock thus providing a clean inner product. Material air locks also allow the sanitisation of products. Tools and other items must be clean and dirt free.

Controls that prevent personnel from the clean area and less clean area being present in the material air lock at the same time. This can be achieved by training and educating staff on the importance of contamination control. A simple visual check of the air lock to confirm it is vacant can be done in order to avoid mixing of personal from different zones. Decontamination procedures are necessary to ensure materials or tools entering the controlled area are decontaminated.

<u>Material Air Lock Considerations:</u>

- Interlocked doors
- Access control
- Sanitation/cleaning procedure
- Double or triple bagged products
- Dedicated trolley for air locks

Disinfection and Cleaning Agents

When materials are being transferred via an air lock, consideration must be given to the status of materials and products. As a rule, no cardboard or unnecessary paper should enter a clean room. Wooden pallets are not acceptable as they can carry dirt and microorganisms and wood cannot be sanitised due to its porous nature. Soft fabric cases often used to carry tools should also be avoided as the material can carry dirt and grease. Cleaning and disinfecting agents should be tested and approved prior to their use onsite. The choice of agents should be backed up with studies that demonstrate the effectiveness of disinfectants and cleaning agents.

Gown-Up Areas

Gowning rooms are designed in order to minimise contamination and facilitate the orderly change over from street clothes to scrubs and/or gowns. Hand washing facilities help reduce the risk of humans carrying unwanted microorganisms into the aseptic processing area. The design of the room should result in clear separation between the less clean side and the clean side. This can be achieved with a step-over segregating the two areas.

Other features of gowning rooms should include:

- Storage lockers for street clothes
- Gown and garment storage
- Body-length mirrors
- Hand washing/drying and disinfection facilities

GMP Zoning

Selecting a suitable classification for a room or manufacturing facility depends on several factors. Firstly, it can be said that sterile products require a more stringent set of criteria than non-sterile products. However, there is an extensive range of products and medical devices that are sterile but are used in different ways and consist of different materials and technology. Some sterile products are single-use only and used for short term purposes and then disposed of. Other sterile products are used subcutaneously for longer periods or even require implantation. Therefore, the design of a facility along with its HVAC specification must be appropriate to the product being manufactured. High-risk products require greater control. The goal of facilities and HVAC systems is to minimise contamination and the associated risks. Using a "sterile versus non-sterile" rule of thumb is not adequate when classifying a room or facility. Standards including EN ISO 14644-1 and guidelines such as EU cGMP Guidelines EudraLex volume 4 Annex 1 (2008) should be consulted in order to fully understand the requirements of each ISO classification and grade of room.

ISO classifications do not specify room occupancy states but when a designation is applied, the occupancy state must be stated in the relevant documentation or procedure. The most relevant European guideline (Annex 1 of the EU cGMP Guideline) lists four classification grades and their associated particulate limits in the 'at-rest' and 'in-operation' conditions. In general, for the sterile and non-sterile products, similar classes are applied, but in non-sterile production the producer could assign their classes, having similar particulate concentration, temperature, pressure etc. but lower air-change rate could be used.

Types of Contamination:

- cross contamination (of a product/material with another product/material)
- non-microbial particulate contamination (non-viable particles)
- biological/microbiological contamination (viable particles/micro-organisms)

Factors Influencing Contamination Cleanliness Levels in the Manufacturing Processes:

- process
- air cleanliness
- personnel hygiene and clothing
- work practices
- material design (material of construction, surface finishes, room finishes, equipment, open system/enclosed system, utensils, etc.)material cleanliness

Room Air Classification (By Limits of Particulate Contamination)

ISO CLASS	FDA	cCMP	Permissible particle number in 1 m3					
			0,1 μm	0,2 μm	0,3 μm	0,5 μm	1 μm	5 μm
1			10	2				
2			100	24	10	4		
3	1		1,000	237	102	35	8	
4	10		10,000	2,370	1,020	352	83	
5	100	A	100,000	23,700	10,200	3,520	832	29
6	1,000	B	1,000,000	237,000	102,000	35,200	8,320	293
7	10,000	C				352,000	83,200	2,930
8	100,000	D				3,520,000	832,000	29,300
9						35,200,000	8,320,000	293,000

Figure 2: Table showing ISO classes and EudraLex Grades A-D.

	Maximum permitted number of particles per m^3 equal to or greater than the tabulated size			
	At rest		In operation	
Grade	0.5 μm	5.0 μm	0.5 μm	5.0 μm
A	3 520	20	3 520	20
B	3 520	29	352 000	2 900
C	352 000	2 900	3 520 000	29 000
D	3 520 000	29 000	Not defined	Not defined

Figure 3: maximum permitted airborne particle concentration for each grade. Showing both "at-rest" and "in-operation" conditions (EU V4 Annex 1). The EU guidance given for the maximum permitted number of particles in the "at-rest" column corresponds approximately to the ISO classifications.

<u>Room Air Classification (By Limits of Microbial Contamination)</u>
The HVAC systems help maintain the viable (microbial) limits within a specific area. These limits are defined in Annex 1 of the EU GMP Guide as shown below.

Grade	Recommended limits for microbial contamination (a)			
	air sample cfu/m^3	settle plates (diameter 90 mm) cfu/4 hours (b)	contact plates (diameter 55 mm) cfu/plate	glove print 5 fingers cfu/glove
A	< 1	< 1	< 1	< 1
B	10	5	5	5
C	100	50	25	-
D	200	100	50	-

Figure 4: Recommended limits for microbial contamination

Environmental Grade A (Aseptic)

Grade A is reserved for critical processes in manufacturing sterile products, product components or product contact. This is generally achieved using isolator technology which maintains a barrier to the background environment or surrounding room.

Grade A Operations include:

- Aseptic processing of sterile ingredients
- Filling of sterile products not for terminal sterilisation
- Stopper insertion
- Crimp capping

Environmental Grade B

Grade B is used for supportive work for aseptic processing corresponding to ISO 14644 (Part 1) Class 5 ("at-rest") and Class 7 (when "in-operation"). Grade B areas typically serve as the background environment of Grade A areas for aseptic processing.

Environmental Grade C

Suitable for non-critical processing steps, Grade C corresponds to ISO 14644 Part 1 Class 7 ("at-rest") and Class 8 ("in-operation"). Grade C operations include:

- Clean side of material air locks and gowning rooms
- Filling of products that are to be terminally sterilised

Environmental Grade D

Grade D at least corresponds to ISO 14644 Part 1 Class 8 ("at-rest" / no definition for "in-operation").

- Clean section of material air locks and final compartments of gowning rooms
- Dispensing of raw materials and excipients and preparation of solutions for sterile products to be sterile filtered and terminally sterilised
- Background environment for transfer and crimp capping of stoppered containers with sterile products

Compliance Tests for GMP Zones

Test	Requirements
Particle count test	Test covers verification of cleanliness. Dust particle counts to be carried out and result printed. The number of readings and positions of tests should be defined in accordance with ISO 14644-1 Annex B5.
Air pressure difference	This test is used to verify non cross-contamination. Log of pressure differential readings to be produced or critical plants should be logged daily, preferably continuously. A 15 Pa pressure differential between different zones is recommended. Refer to ISO 14644-3 Annex B5.
Airflow volume	To verify air change rates. Airflow readings for supply air and return air grilles to be measured and air change rates to be calculated. Refer to ISO 14644-3 Annex B13.
Airflow velocity	To verify unidirectional flow or containment conditions. Air velocities for containment systems and unidirectional flow protection systems to be measured. Refer to ISO 14644-3 Annex B4.
Filter leakage tests	To verify filter integrity. Filter penetration tests to be carried out by a competent person to demonstrate filter media, filter seal and filter frame integrity. Only required on HEPA filters. Refer to ISO 14644-3 Annex B6.
Containment leakage	To verify absence of cross-contamination. Demonstrate that contaminant is maintained within a room by means of: • airflow direction smoke tests • room air pressures. Refer to ISO 14644-3 Annex B4.
Recovery	To verify clean-up time. Test to establish time that a clean room takes to recover from a contaminated condition to the specified clean room condition. Should not take more than 15 minutes. Refer to ISO 14644-3 Annex B13.
Airflow visualisation	To verify required airflow patterns. Tests to demonstrate air flows: • from clean to dirty areas • do not cause cross-contamination • uniformly from unidirectional airflow units Demonstrated by actual or video-taped smoke tests. Refer to ISO 14644-3 Annex B7.

Environmental Monitoring

An environmental monitoring programme is required for GMP-controlled areas. The purpose of such programmes is to document, define and describe parameters to be monitored, monitoring both frequency and methods. Environmental monitoring is a regulatory requirement. It also demonstrates that the GMP areas are being controlled and are fit for purpose.

Key Requirements of Environmental Monitoring

- Identification and classification of environmental areas that require monitoring
- Test methods and sampling procedures
- Defined testing frequencies
- Sample locations based on Risk
- Microbial monitoring of personnel
- Monitoring of non viable particles
- Monitoring of temperature, relative humidity and differential pressures
- Defined alert and action levels for each environmental area
- Trending of Enviromental data
- Change Control

Other parameters such as those controlled by the HVAC system (air changes/hour etc.) should also be verified according to a defined schedule.

<u>Grade A, B and C</u>

- Viable and non-viable particles monitored under operational conditions
- Risk-based approach to sampling points and represent high-risk/critical positions

Grade D

- Non-viable particles must be measured in at-rest conditions
- Viable particles measured under operational conditions

Further reading

ISO 14644-1: International Organisation For Standardisation – Cleanrooms and Associated Controlled Environments. Part 1: Classification of Air Cleanliness.

ISO 14644-3: International Organisation For Standardisation – Cleanrooms and Associated Controlled Environments. Part 3: Test Methods.

ISO 14644-4: International Organisation For Standardisation Cleanrooms and Associated Controlled Environments: Part 4: Design, Construction and Start-Up.

EudraLex, Vol 4, Annex 1: EU Guide to Good Manufacturing Practice (EU GGMP) Governing Medicinal Products for Human and Veterinary Use, Annex 1 – Manufacture of Sterile Medicinal Products.

EN 1822:2009: European Standard For HEPA Filter Classification.

US FDA CFR 211: Code of Federal Regulations Food and Drug Administration Title 21 Part 211 – Current Good Manufacturing Practice for Finished Pharmaceuticals – Section 211.46 Ventilation, Air Filtration, Air Heating and Cooling.

ICH Q7: International Conference on Harmonisation - Good Manufacturing Practice Guide for active Pharmaceutical Ingredients – Section 4.21 and 4.22 – Utilities.

US FDA: Food and Drug Administration - Guidance for Industry "Sterile Drug Products Produced by Aseptic Processing – Current Good Manufacturing Practice".

CHAPTER 2

CLEAN UTILITIES

Introduction

The term "clean utilities" in the life science industry refers to utilities that have to fulfil regulatory requirements. The most common utility is water, which can be supplied in different pharmaceutical grades of purity. Purified water (PW or PUW), highly purified water (HPW) and water for injection (WFI) are the most common. Water quality specifications can be found in the pharmacopeias, e.g. the US Pharmacopeia. Other clean utilities can also include clean compressed air, clean gases (e.g. nitrogen, argon and oxygen), and clean steam.

Key Definitions

Alert Limit: a value reached when the normal operating range of a critical parameter has been exceeded, indicating that corrective measures may need to be taken to prevent the action limit being reached.

At-Rest: a condition where the installation is complete with equipment installed and operating in a manner agreed upon by the customer and supplier, but with no personnel present.

Clean Room: an area (or room or zone) with defined environmental control of particulate and microbial contamination, constructed and used in such a way as to reduce the introduction, generation and retention of contaminants within the area.

Containment: a process or device to contain product, dust or contaminants in one zone, preventing it from escaping to another zone.

Contamination: the undesired introduction of impurities of a chemical or microbial nature, or of foreign matter, into or onto a starting material or intermediate, during production, sampling, packaging or repackaging, storage or transport.

Point Extraction: air extraction to remove dust with the extraction point located as close as possible to the source of the dust.

Pressure Cascade: a process whereby air flows from one area, which is maintained at a higher pressure, to another area at a lower pressure.

Relative Humidity: the ratio of the actual water vapour pressure of the air to the saturated water vapour pressure of the air at the same temperature expressed as a percentage. More simply put, it is the ratio of the mass of moisture in the air, relative to the mass at 100% moisture saturation, at a given temperature.

Turbulent Flow: turbulent flow, or non-unidirectional airflow, is air distribution that is introduced into the controlled space and then mixes with room air by means of induction.

Identifying Critical Utilities

The process of identifying critical utilities can be done with the application of direct impact, indirect impact and no impact definitions (see previous section *"risk and impact assessment"*). Risk assessments, CQAs and CPPs should also help identify critical utilities. When critical utilities are required as part of manufacturing and processing, the following points should be examined during the requirements and design stage:

- Materials of construction
- Internal surface finishes
- System sizing
- Flow rates, dead legs, drainage etc.

The process of identifying critical utilities can be done with the application of direct impact, indirect impact and no impact definitions (see previous chapter). Risk assessments, CQAs and CPPs should also help identify critical utilities. When critical utilities are required as part of manufacturing and processing, the following points should be examined during the requirements and design stage:

- Materials of construction
- Internal surface finishes
- System sizing
- Flow rates, dead legs, drainage etc.

Compressed Air

Compressed air is used for valve actuation, instrument air and process air to name but a few applications. Only the point-of-use filtration and the gas quality instrumentation should be classified as level 1. When flow or pressure is a CPP, the measurement/monitoring should be performed by the system into which the gas is flowing. Additionally, the CQAs and CPPs should be routinely monitored through the calibrated monitoring system. For compressed air, the potential CPPs are listed below. For the physical system being evaluated, the use and the application of the compressed air will determine which (if not all) CPPs are needed to ensure the system produces product of the desired quality.

- Hydrocarbons
- Moisture
- Particulates
- Temperature

It is important that each point of use has appropriate sterile filters in place. If the filter is not placed directly at the point of use, control and counter measures should be implemented to address any risk of contamination downstream of the filter. Compressed air for bio-pharmaceutical use must be generated using oil-free compressors with appropriate temperature controls in place.

Water Systems

Water supply and the associated water systems in biotechnology and pharmaceuticals are vital components of the manufacturing process. They are used to clean equipment and vessels, to cool or heat processing pipes and systems, and in many circumstances certain grades of water are components of the finished product (e.g. water-for-injection). Various grades of water service a particular purpose. Some common types include:

- Potable water
- Soft water
- Purified water
- Water-for injection

Water used in-process and in-cleaning should be pure and free from microbial and chemical impurities. As the water gets easily contaminated by environmental conditions, diligence in the design is essential. Typically water systems are supplied on a continuous loop with recirculation.

CPPs typical for a water system include:

- Pressure
- pH
- Conductivity
- Level
- TOC
- Flow
- Temperature
- Resistivity

Water-for-Injection

The use of WFI is twofold. Firstly, it can be used for critical processing steps such as washing and rinsing . It can also be used in injectable products. WFI is a key raw material for sterile intravenous and intradermal products. WFI is produced by Multi Column Distillation Plant (MCDP), and must meet the microbial requirements of regulated bodies.

Clean-in-Place (CIP) / Sterilise-in-Place (SIP) System

The cleaning of equipment, vessels and process piping is a critical activity. Any residue from a previous production batch needs to be removed in order to avoid cross contamination. CIP and SIP skids are often utilised to allow efficient switchover between batches and/or products.

Clean steam

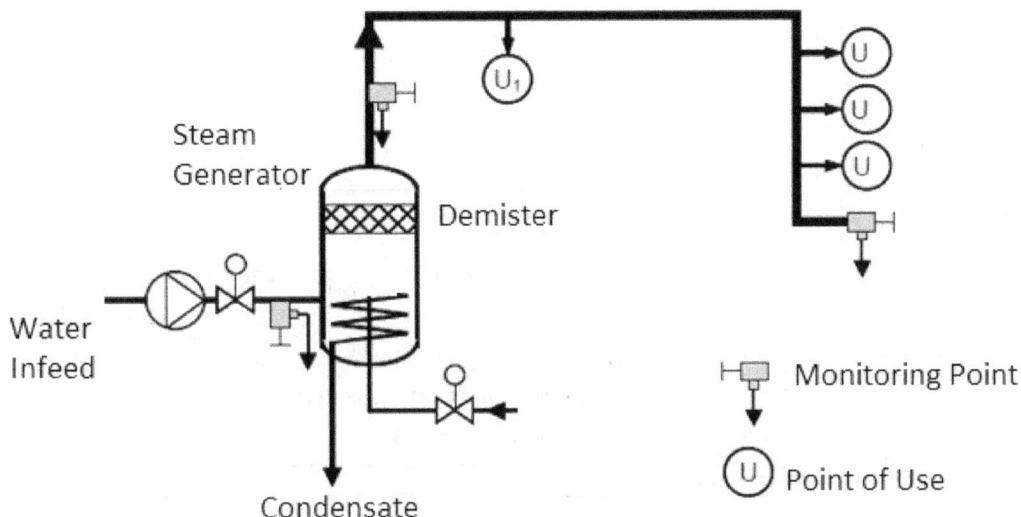

Figure 5: Simple Clean Steam Generation Piping and Instrumentation

Pure Steam is used in pharma and biotech for sterile application, autoclave sterilisation etc. Distribution piping of clean steam is a critical aspect. Improper sizing of pipes may impact the production process and lead to loss of time during sterilisation.

Clean steam, also referred to as "pure steam", and gases used in manufacturing operations must be of a quality suitable for their intended purpose. The intended use of clean steam and gases must be understood in order to determine any risks to the patient or product. For example, gases that end up being part of the product must fulfil the regulatory requirements. Preventative maintenance and on-going monitoring must be implemented for clean steam systems.

- ➢ Routine inspection and maintenance
- ➢ Frequency of filter change
- ➢ Frequency of the sterilisation for the gas distribution system, if applicable
- ➢ Frequency for integrity testing of the sterile filter

Water systems for purified water, de-ionised water and water-for-injection (WFI) must provide a consistent and reproducible output. Where there is moisture, there is always a risk of microbial contamination. Therefore, the design of water systems should mitigate against such risks. Good engineering practices such as using circulation loops, no dead legs and polished surface finishes all work to provide an effective and safe system. The design should also take into account ease of sampling at the point of use. The removal of endotoxins is a requirement for WFI.

On-going sampling to monitor the quality of water is particularly important where water systems are concerned. Procedures should be in place to ensure effective monitoring and testing is maintained. Action limits and acceptance criteria should be clearly documented in approved SOPs or equivalent. Failure to meet limits or acceptance criteria should initiate an investigation. The potential CPPs are listed below for clean steam systems:

- Conductivity
- Flow
- Level
- Pressure
- Resistivity
- Temperature

Design Considerations

The purpose of a User Requirement Specification (URS) is to define the requirements for the operation and control of the clean steam system.

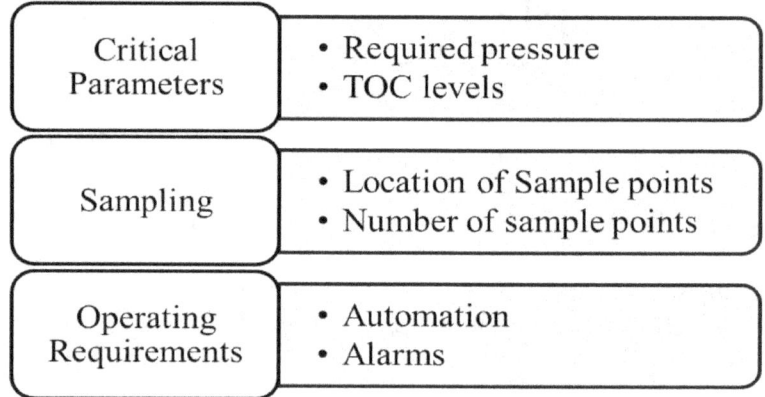

HVAC

Heating, ventilation and air-conditioning (HVAC) plays an important role in ensuring the manufacture of quality products. Furthermore, HVAC systems also provide comfortable conditions for operators based in the manufacturing environment. HVAC system design influences the layout of airlock positions and doorways. In turn, airlocks, entrances and exits have an effect on room pressure differential cascades and cross-contamination control. The prevention of contamination and cross-contamination is an essential design consideration of the HVAC system. In view of these critical aspects, the design of the HVAC system should be considered at the concept design stage of a manufacturing plant.

Temperature, relative humidity (RH) and ventilation should not adversely affect the quality of products during their manufacture and storage, or the proper functioning of equipment. CPPs for HVAC systems include:

- Temperature
- Humidity
- Particle count (viable and non-viable)
- HEPA filter certification/leak test/air flow rates
- Room differential pressures

The Displacement Concept (Low Pressure Differential, High Airflow)

This concept is commonly found in production processes where large amounts of dust are generated. Under this concept the air should be supplied to the corridor, flow through the doorway, and be extracted from the back of the cubicle. Normally the cubicle door should be closed and the air should enter the cubicle through a door grille, although the concept can be applied to an opening without a door. The velocity should be high enough to prevent turbulence within the doorway resulting in dust escaping. This displacement airflow should be calculated as the product of the door area and the velocity, which generally results in relatively large air quantities.

Note: This method of containment is not the preferred method as the measurement and monitoring of airflow velocities in doorways is difficult.

Pressure Differential Concept (High Pressure Differential, Low Airflow)

The pressure differential concept may normally be used in zones where little or no dust is being generated. It may be used alone or in combination with other containment control such as a double door airlock. The high pressure differential between the clean and less clean zones should be generated by leakage through the gaps of the closed doors to the cubicle. The pressure differential should be of sufficient magnitude to ensure containment and prevention of flow reversal, but should not be so high as to create turbulence problems.

In considering room pressure differentials, transient variations, such as machine extract systems, should be taken into consideration. A pressure differential of 15 Pa is often used for achieving containment between two adjacent zones, but pressure differentials of between 5 Pa and 20 Pa may be acceptable. Where the design pressure differential is too low and tolerances are at opposite extremities, a flow reversal can take place. For example, where a control tolerance of $\pm$ 3 Pa is specified, the implications of rooms being operated at the upper and lower tolerances should be evaluated. It is important to select pressures and tolerances such that a flow reversal is unlikely to occur. The pressure differential between adjacent rooms could be considered a critical parameter, depending on the outcome of risk analysis.

The limits for the pressure differential between adjacent areas should be such that there is no risk of overlap in the acceptable operating range, e.g. 5 Pa to 15 Pa in one room and 15 Pa to 30 Pa in an adjacent room, resulting in the failure of the pressure cascade, where the first room is at the maximum pressure limit and the second room is at its minimum pressure limit. Low pressure differentials may be acceptable when airlocks (pressure sinks or pressure bubbles) are used to segregate areas.

The pressure control and monitoring devices used should be calibrated and qualified. Compliance with specifications should be regularly verified and the results recorded. Pressure control devices should be linked to an alarm system set according to the levels determined by a risk analysis. Manual control systems, where used, should be set up during commissioning, with set points marked, and should not change unless other system conditions change. Airlocks can be important components in setting up and maintaining pressure cascade systems and also help to limit cross-contamination. Airlocks with different pressure cascade regimes include the cascade airlock, sink airlock and bubble airlock.

CHAPTER 3

STERILE MANUFACTURING OPERATIONS

Introduction

Sterile manufacturing operations depends on several factors including the right design and operation of facilities, utilities and equipment. Sterility assurance must be demonstrated to be in control within a manufacturing setting. This is achieved by:

- Qualification and validation of the processes, facilities, utilities, equipment, cleaning methods and sterilisation operations
- Qualified personnel for aseptic handling in conventional clean rooms or by barrier systems
- Control of critical aspects and critical parameters via the application of change management, change control and a suitable quality management system
- Environmental monitoring
- Routine Maintenance
- Analytical method validation

Sterility

The impact of contaminated injectable products can result in serious illness or death to patients. Many injectable treatments sustain life and bio-chemical processes or genetic conditions. While there is always residual risks or acceptable risks, it is important to mitigate against any risks throughout the manufacturing process. Furthermore, a risk-based approach to operations and in particular changes to the process must be maintained throughout the life cycle of a product. Contamination can be caused by particles or microbes.

Where appropriate and technically permissible terminal sterilisation is the preferred point of sterilisation. Terminal sterilisation is when the final sealed product in its container is sterilised at the end of the process.

Unit Operations

Bioprocesses treat raw materials and generate useful products. Unit operations are the individual steps in the process that modify materials and their properties at each step of the process. Each unit operation comes together to create a complete process. The term unit operation usually refers to processes that cause physical modifications to materials such as a change in phase or component concentration. Chemical or biochemical changes are the subject of reaction engineering.

Bioreactor Engineering

The design and manufacture of bioreactors is yet again an area within bioprocessing that depends on scientific and engineering expertise. It should be pointed out that there is no standard design procedure for the design of reactors. However, knowledge of bioprocess reactions and kinetics is a key element. Other knowledge such as mixing, mass transfer and heat transfer also contribute to the design process. Key aspects of bioreactor design include:

Reactor size: What is the capacity of the reactor? This is generally driven by the expected production volumes.

Reactor configuration: Is the reactor air driven, stirred, agitated etc.?

Operating configuration: Is it a continuous operation or a batch driven operation?

Process Requirements: Refer to the required operating temperatures, pH that needs to be maintained in the vessel.

Stirred Tank

A conventional tank involves mixing and bubble dispersion done via mechanical agitation. This requires a high energy input per unit volume.

Headspace is an important consideration when filling tanks. Typically, only between 60% to 80% of the tank volume is used. This headspace is important especially if foaming of the broth occurs.

Some tanks are designed to take account of foaming issues with the addition of a foam. Chemical means of reducing or preventing foam formation can also be employed. However, these chemicals can impact the process (reduction in rate of oxygen transfer). Temperature modulation is typically controlled using coils.

Bubble Column Bioreactor

A bubble column is a type of bioreactor. Bubble columns offer an alternative to stirred reactors, having no mechanical means of stirring. Mixing and aeration is done by gas sparging by the use of a gas sparger placed at the bottom end of the vessel. This type of reactor requires a lot less energy to mix compared to mechanical stirring.

Airlift Reactors

Airlift reactors are similar to bubble columns as neither require mechanical mixing. A key difference between bubble columns and airlifts is that the air is channelled through a riser in the airlift, which allows more control of the bubble patterns. Airlift reactors can be categorised into either internal loop or external loop configurations.

Aseptic Operation

With the exception of food and beverage fermentations, cultures used in the treatment of medical conditions frequently require sterile conditions.

This is especially important for slow growing cultures that can be quickly compromised by unwanted contaminates. Typically, up to 5% of fermentations in industrial settings are lost as a result of failings in sterilisation. Slow growing cells would have a higher rate of contamination due to sterility issues. Antibiotics by their nature have a higher resilience to this type of loss.

Industrial fermenters are designed to allow in-place steam sterilisation under pressure.

For effective steam sterilisation, the vessel must be fully purged of air. Dead legs, stagnant areas or crevices should be avoided during the design phase as these can be a point of microbial contamination. Polished welded joints with a high surface finish are desired.

Valves

Valves control the introduction of liquids to the vessel and their removal when required. Valves therefore, can be a potential entry point for contaminants. Traditional gate and globe valves do not suffice for aseptic operations.

Pinch and diaphragm type valves are more commonly used as they do not contain any dead spaces within their assembly. The closing mechanics also provide isolation from the liquid or product contents.

Materials of Construction (MOC)

Fermenters are made of materials that are suited to the use of steam sterilisation techniques and regular cleaning. These materials can be classed as both non-reactive and non-absorptive surfaces. Most large-scale reactors are made of high-grade stainless steel. Cheaper classifications of stainless steel can be used for jacketing and other non-product contact areas.

All interior product contact surfaces should be polished to a "mirror" finish. Welds also need to be finished in a similar manner. Electro polishing provides a better quality surface finish than mechanical polishing.

As with any chemical reaction, factors such as temperature, pH and oxygen concentration can impact the performance and yield. To ensure the optimum conditions are maintained, it is important to monitor and control such parameters and factors. By far the most common these days is automatic control of systems and equipment with automatic feedback and adjustment.

Downstream Processing

In fermentation processes, raw materials are altered most significantly by the reactions occurring in the fermenter. In addition, physical changes after fermentation are also important in order to extract and purify the desired product from the culture broth. Any treatment completed after fermentation is referred to as downstream processing. In most instances, downstream processing only requires physical modification of material rather than any chemical or biochemical processing. Although the product will dictate the downstream processing model, there are general major steps which are detailed below.

CELL REMOVAL: this step involves the removal of cells from the fermentation liquor. If the cells are the product itself, little downstream processing is required. Typical unit operations for cell removal include filtration, microfiltration and centrifugation.

CELL DISRUPTION/CELL DEBRIS REMOVAL: If the product desired is located within the cell itself, then these unit operations are required in order to open the cells and release their contents. Such a unit operation includes high-pressure homogenisation. The cell debris is then separated from the desired product via filtration unit operations.

PRIMARY ISOLATION: The purpose of primary isolation is to remove components that differ from the product. Unit operations such as solvent extraction, precipitation and ultrafiltration are typically used.

PRODUCT ENRICHMENT: The purpose of product enrichment is to separate the product from impurities with properties that are close to those of the product. Chromatography is usually used at this stage of the process.

FINAL ISOLATION: The method used for final isolation depends greatly on the product in question. Ultrafiltration is used for liquids and drying for solid products.

Removal Operations

One of the first steps in downstream processing is the removal of cells from the culture liquid. The major process options for cell removal are filtration, microfiltration and centrifugation.

In general, filtration and microfiltration use particle size as the principle of operation, whereas centrifugation relies on particle density. Other factors such as viscosity of broth and surface charge contribute to the performance of these removal operations.

Filtration

Basic filter design involves solids being retained in the filter cloth, while the liquid passes through the cloth/membrane. However, the liquid filtrate that passes through typically contains a small portion of solids. It should be noted that large scale filtration is expensive and difficult to perform under sterile conditions.

Microfiltration

Microfiltration uses microporous membranes to recover cells. Unlike filtration, microfiltration generally does not require preconditioning (heating or addition of agents to reduce viscosity). Microfiltration allows cell recovery of typically 100%, so it is therefore very efficient. Microfiltration can also be done under sterile conditions. It is also typically less expensive than filtration and centrifugation.

Centrifugation

Centrifugation is ideal for cell recovery if the cells or product is too small to filter using conventional filtration. Centrifugation of fermentation broths results in a thick cream-like sludge. Another advantage of centrifugation systems is that many of them are steam-sterilisable. Centrifugation must always take place under sterilised conditions.

Precipitation

Precipitation is a method that is frequently used for the recovery of proteins from culture broths. It also has applications in downstream processing for products such as antibiotics. Typically, precipitation is used in the early steps of downstream processing as it facilitates the reduction in liquid volume, which therefore makes the preceding steps less costly and easier to manage. Precipitation is achieved by the adding of *precipitants* such as salts, solvents and polymers to work to change properties such as pH, ionic strength or temperature of the solution. These effects reduce the solubility of the product which forces it to precipitate out of the solution in particles which are insoluble. The precipitated solid particles can then be recovered by filtration, microfiltration or centrifugation.

Proteins and Precipitation

The most common application of precipitation is recovery of proteins. Proteins treated using precipitation include enzymes used for medical treatments, food proteins from plants and animals and also recombinant proteins manufactured using genetically engineered organisms.

Precipitation Methods

Examples of precipitation methods include:

- Salting out
- Isoelectric precipitation
- Organic solvent precipitation

In summary, the goal of precipitation is to reduce solubility of materials, hence inducing precipitation, which forces solid formation (particles of product). For protein recovery, precipitation is aimed at separating the protein from solution, without causing damage or changes that cannot be reversed.

Salting Out

High salt concentrations facilitate the aggregation and precipitation of proteins. The salt causes the water surrounding the protein to move into the bulk solution. This creates a "hydrophobic zone" on the protein surface. In simple terms, "hydrophobic" means water repelling — a surface that does not take up water due to the hydrophobic zones allowing sites of attraction between the protein molecule and other proteins within the solution. The success and applicability of salting out depends on the hydrophobicity of the protein. Proteins that have few hydrophobic zones tend to remain in solution — even at high salt concentrations.

Isoelectric Precipitation

Isoelectric precipitation works by reducing the electrostatic forces to near zero, allowing the proteins to precipitate out. A benefit of isoelectric precipitation compared to salting out is that desalting of the precipitate is not required.

Organic Solvent Precipitation

The addition of solvents such as ethanol or acetone to aqueous protein solutions generally causes the protein to precipitate. Organic solvents have a lower dielectric constant than water which means they store less electrostatic energy when compared to water. Therefore, in the presence of the solvent, oppositely charged groups of proteins experience greater attractive forces which cause protein aggregation and precipitation.

Membrane Filtration

Membrane filtration is another type of unit operation that is used in downstream processing. It can be applied in order to separate, concentrate or purify a product. Applications include:

- Cell removal
- Cell debris removal
- Desalting
- Removal of viruses
- Recovery of precipitates

Membrane filtration has a number of advantages compared to other unit operations used to concentrate products:

- Low process energy requirements
- Membrane filtration can be done aseptically
- Does not need harsh chemicals

Membrane filtration can be categorised according to the size of the particles that are retained by the membrane:

Microfiltration: used to remove particulate such as cells and cell debris ranging in size from 0.2 to 10μm from broths. Typical membranes have a nominal pore size diameter of 0.05 to 5μm.

Ultrafiltration: Membranes for ultrafiltration have pores typically of a nominal size between 0.001μm to 0.1μm.

Raw Materials

Raw materials and components used in the manufacturing process should be properly sourced and approved through a supplier quality programme. This ensures that the vendor or supplier of raw materials has the necessary regulatory status and quality controls in place. A robust supplier approval process ensures materials are provided consistently to pre-approved specifications. Raw materials for sterile products must be tested for their bioburden and when necessary for bacterial endotoxin levels to determine acceptability of their use.

Upstream Processing

Most aseptic filling processes are made up of a number of key steps. However, it must be noted that the compounding processes required in order to supply the active product used in filling can be complex in nature. Filling and closing operations tend to be more similar in nature across different companies and different products.

The first step normally involves some format of a glass container such as a vial or bottle being processed through a washer. Utilising ultrasonics, heated WFI baths, WFI spray and process air blowing, components are washed to remove any dirt or debris. At the outfeed of the washer, components then travel into a depyrogenation tunnel where they are dried, depyrogenated and sterilised.

While the washing of containers and vials is an important step in many manufacturing processes, it does not clean or sterilise components. Its role is to remove any particles or debris. Typically no detergents are used for vials destined to deliver intravenous products. As no detergents or other chemicals are used, cleaning does not occur. The term "cleaning" applied to the biopharmaceutical industry refers to the removal of soils or greases. Vial washers using ultrasonic systems, heated WFI and process air blowers are not designed to "clean" in this regard.

Filling Operations

Suspensions and solutions that are filled in glassware such as vials provide lifesaving and sustaining medical treatments for millions of patients worldwide. When the product reaches the filling unit operation, it has been through many unit operations. The product and components must be sterile at this point. Transfer of product to individual vials or containers may be facilitated by employing piston valves, pressure control and peristaltic pumps.

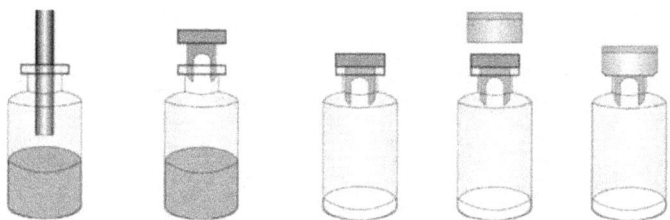

Figure 6 : Representation of filling and closing of vials.

Once the required quantity of solution or suspension has been filled, the next unit operation required is container closure achieved by the insertion or application of a stopper or cap. Key consideration for filling and closing operations include:

> ➢ Design and function of filler heads
> ➢ Design and function of filler needles

> Fill accuracy and fill weight

The filling of Biotechnology Derived Products (BDP) into ampules or glass vials presents similar problems as with the processing of conventional products. Attempting to develop a site, prove clinical effectiveness and safety, as well as the validation of sterile operations, equipment, processes and systems often necessitates a lengthy process to achieve success for a start-up BDP facility.

The batch size initially produced by a BDP is likely to be small. Because of the small batch size, filling lines may not be as automated as for other products typically filled in larger quantities. Thus, there is more involvement of people filling these products. This can present more chances of contamination meaning any operation or involvement must be controlled and monitored.

Problems that have been identified during filling include inadequate attire, deficient environmental monitoring programmes, hand-stoppering of vials, particularly those that are to be lyophilised and failure to validate some of the basic sterilisation processes.

Because of the active involvement of people in filling and aseptic manipulations, the number of persons involved in these operations should be minimised, and an environmental programme should include an evaluation of microbiological samples taken from people working in aseptic processing areas.

Another concern about product stability is the use of inert gas to displace oxygen during both the processing and filling of the solution. As with other products that may be sensitive to oxidation, limits for dissolved oxygen levels for the solution should be established. Likewise, validation of the filling operation should include parameters such as line speed and location of filling syringes with respect to closure, to ensure minimal exposure to air (oxygen) for oxygen-sensitive products.

In the absence of inert gas displacement, the manufacturer should be able to demonstrate that the product is not affected by oxygen.

Typically, vials to be lyophilised are partially stoppered by machine. However, some filling lines have been observed that utilise an operator to place each stopper on top of the vial by hand. The concern is the immediate avenue of contamination offered by the operator. The observation of operators and active review of filling operations should be performed.

Another major concern with the filling operation of a lyophilised product is assurance of fill volumes. A low fill would represent a sub-potency in the vial. Unlike a powder or liquid fill, a low fill would not be readily apparent after lyophilisation, particularly for a product where the active ingredient may be only a milligram. Because of the clinical significance, sub-potency in a vial can potentially be a very serious situation.

A common method of filling vials consists of a two-step filling process. Generally, the first step fills up to 90% of the vial, with the second more accurately filling the remaining amount.

The following parameters must be maintained to achieve the same fill volumes at each filling cycle:

> Viscosity of the product
> Product temperature
> Pressure in the dosing vessel
> Level in the dosing vessel
> Needle/filling head properties

- Properties of the hose material

Container Closure Integrity

Upstream processes need to take into account the many requirements that aim to produce products that are safe and effective for patients. Operations such as dispensing and compounding apply GMP principles from the very beginning of the manufacturing process. When the product has been manufactured and is ready to be filled and closed, so too the container closure methods must ensure that sterility and integrity of the product is preserved. Therefore, sterile product container closure systems (or closing systems) must be designed, qualified, and controlled in accordance with international and local regulatory requirements and GMP guidance.

<u>Regulatory Requirements</u>

FDA 21 CFR Part 600. PART 600 -- BIOLOGICAL PRODUCTS: GENERAL Subpart B--Establishment Standards, h)

h) Containers and closures.

"All final containers and closures shall be made of material that will not hasten the deterioration of the product or otherwise render it less suitable for the intended use. All final containers and closures shall be clean and free of surface solids, leachable contaminants and other materials that will hasten the deterioration of the product or otherwise render it less suitable for the intended use. After filling, sealing shall be performed in a manner that will maintain the integrity of the product during the dating period. In addition, final containers and closures for products intended for use by injection shall be sterile and free from pyrogens. Except as otherwise provided in the regulations of this subchapter, final containers for products intended for use by injection shall be colourless and sufficiently transparent to permit visual examination of the contents under normal light. As soon as possible after filling final containers shall be labelled as prescribed in 610.60 et seq. of this chapter, except that final containers may be stored without such prescribed labelling provided they are stored in a sealed receptacle labelled both inside and outside with at least the name of the product, the lot number, and the filling identification."

<u>Quality Requirements</u>

Product containers and closure systems must be capable of being sterilised and depyrogenated before the product is filled. A simple example would be glass vial and stopper components used for various injectable products undergoing sterilisation as part of the process (depyrogenation and sterilisation tunnels for glass components, and stopper sterilisation using vessels and moist heat).

During the product development stage, the type of container closure system must be developed based on the intended use, physical and chemical requirements of the medicine, storage requirements, delivery methods and shelf life. A detailed testing strategy should be developed as early as possible to test for the suitability of the container components. Test strategies can be best developed using a risk-based approach along with knowledge and experience of personnel. In addition to the components used and the size and shape of components, engineering studies are also required in order to define the critical parameters to be used during container closing operations. Depending on the methods of closure, some critical parameters may include:

- Crimping force (in crimp caps are used)

- Closure torque
- Stopper position
- Stopper force
- Closure

The parameter selection must ensure the integrity of the container closure system is not compromised. Principles of quality management must be applied to container closure systems when validated and should address the following:

- Approved container closure components
- Control srategy for critical parameters
- Finished product release testing
- Changes to the container closure system managed under formal change control procedures

Testing

Integrity testing is the most critical test with regard to closed containers. Depending on the container closure format, the followings tests may be used:

- Dye bath test
- Vacuum test
- Headspace analysis
- Electronic spark test
- Bubble testing
- Leak testing
- Pressure decay
- 100% visual inspection
- Presence of stopper or closure cap
- Presence of tip/cap
- Presence of cracks, sealing issues or evidence of crimping deficiencies e.g. crimp height.

Electronic Spark Test

This test allows the detection of very small pinholes or cracks that may cause leaks or effect the container integrity. It is used to assess ampoules, vials and glass cartridges. By placing high-voltage through the item and measuring impedance, any cracks or closure issues can be identified in samples.

Vacuum Test

This is completed by applying a vacuum for a defined time and then allowing it to reach ambient pressure. This is a common test used, however it can lack sensitivity.

Head Space Analysis

For containers that are filled using nitrogen or some other inert gases such as argon, head space analysis provides an accurate and repeatable test method. The O2 levels can be detected by the increase in oxygen content. A non-destructive way of testing for head space analysis is to use laser measurement.

Aseptic Process Simulation

Validation of aseptic processing for products must include simulating the process using aseptic process simulations. For simulations of final product filling, the number of containers filled should be representative of the projected batch size and be sufficient to enable a valid evaluation, including all routine operator interventions.

Isolator Barrier Systems

An isolator is a complex barrier system designed to support aseptic processing and manufacturing. The supplied air to such systems is generally supplied through a microbially-retentive filtration system. High efficiency particulate air (HEPA) filters are capable of removing particles as small as 0.3μm making them an integral part of isolator technology.

Figure 7: Photograph of a typical isolator showing isolator doors fillted with glove ports.

HEPA filters should be capable of achieving Grade A (ISO Class 4.8) at-rest and in-operation. Some exceptions are permitted, such as powder filling, however, risk assessments should mitigate risk to patients. The isolator is a sealed enclosure where there is no direct opening to the external environment or room. Transfer of materials or utensils is done in a controlled manner using a decontaminated interface.

Isolator Interfaces

Depending on the design considerations and individual vendor designs, isolators can have a number of operation interfaces. The term "interface" refers to the ability of an operator or process technician to interact with the machine. The primary method of intervention utilizes glove systems.

Four part glove systems consisting of a gauntlet, glove, cuff-ring and sleeve. When used properly and by trained personnel, glove systems support critical line interventions required during aseptic processing and manufacturing.

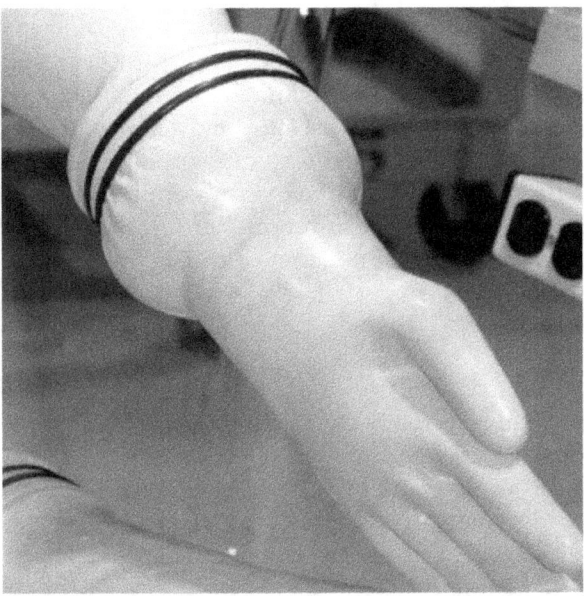

Figure 8: Isolator glove inflated (undergoing glove integrity test).

Figure 9: Rapid Transfer Port (RTP). Another means of transferring materials and tools.

The gasket of RTP systems has been identified as a potential source of contamination in isolators since there may be a small contact surface around the gasket that may not be exposed to the decontaminating agent. A risk analysis should be done to evaluate potential contamination risks with the gaskets and the need for maintenance programmes. Transfer of material into and out of the isolator is also a potential source of contamination.

Furthermore, isolators may also be designed in combination with smaller enclosures associated with them to allow the continuous ingress of materials through the smaller isolator into a main isolator.

Classification of Isolator Rooms

The surrounding room of an isolator should have limited access to staff (ensuring only the presence of authorised personnel), adequate space around the isolator and temperature/humidity under control for the effective utilisation of decontamination technologies (e.g. vapour phase hydrogen peroxide systems).
Regulatory authorities require background environments of aseptic production isolators to be classified at minimum in zone (Grade) D (ISO 8 at-rest). However, there is a general consensus that sterility testing isolators need not be placed in a classified clean room, but it is important that such isolator surrounding rooms impose restricted access.

Isolator Decontamination

The purpose of bio-decontamination is to remove viable bioburden on exposed surfaces inside the isolator; a decontamination process should be performed using sporicidal chemical agents associated with decontamination equipment such as gas/vapour phase decontamination systems using hydrogen peroxide (e.g. VHP) or the equivalent. A decontamination cycle is an automated machine cycle that is controlled and monitored during each stage of the cycle. Cycles can be divided into four stages:

- Dehumidification
- Conditioning
- Decontamination
- Aeration

Dehumidification: The dehumidification stage (also known as pre-conditioning) is designed to ensure that the isolator enclosure has a predefined humidity value (< 20 % RH) to ensure a proper concentration of decontaminating agent.

Conditioning: Depending on the complexity of the system, at a minimum, the isolator must have a tightly controlled temperature range, positive pressure and air velocity control. During this initial stage, the isolator doors and ports must be closed and sealed. Any defects in the barrier system should result in an alarm and abort the cycle. During conditioning, an automated leak test should be initiated to detect any breaks in the barrier system (e.g. defective gloves or seals). Heating of VHP delivery pipework also occurs. The conditioning stage is when the decontaminating agent shall reach the minimum concentration required to achieve the desired microbial reduction.

Decontamination: At this stage the VHP is maintained in the isolator according to the dosing rate contained in the recipe or cycle settings.

The time and total amount of VHP must result in a kill in BIs placed within the isolator. Generally a 6 log reduction is required for a cycle to be deemed a success.

Aeration: During the aeration stage the amount of residual decontaminating agent must fall to safe levels. (< 1ppm). This is done by blowing the hydrogen peroxide carrying air out of the barrier system using fresh air.

Recommended Critical Process Parameters	Typical Units
Amount of H2O2 during conditioning	(g)
Dosing rate (conditioning)	(g/min)
Time for conditioning	(mins)
Amount of H2O2 during decontamination	(g)
Dosing rate decontamination	(g/min)
Time for decontamination	(mins)
Aeration time	(mins)

Decontamination Agents

Decontamination of isolators is achieved by the supply of gaseous sporicidal agents. These agents must be capable of killing both bacterial endospores and fungal spores. The system typically turns liquid agents into a gaseous vapour.

The decontamination agent typically used in industry is hydrogen peroxide. Other agents include formaldehyde, peracetic acid and chlorine dioxide. The rationale for selecting a particular agent should be based on technical data, sporicidal efficacy and the materials and products that come into contact with such agents.

Often the starting point when selecting an agent is the manufacturer's recommendations. Manufacturers of equipment trains are best positioned to understand interactions with seals and surfaces etc. In many cases, the equipment is designed with a particular type of decontamination agent in mind.

Another source of information is the datasheets provided by the agent manufacturers. Datasheets also give an insight into the suitability of a chemical based on its purity, concentration and safety.

The below factors should be considered with regards to biodecontamination:

- Ensure as much surface area as possible of components are exposed.
- Minimise loads in order to limit the bioburden levels prior to the cycle starting.
- For filling and closing machines, design automation to ensure parts are moving during the cycle to facilitate exposure to the agent.
- Ensure all areas are dry and free of foreign objects and debris.

Containment Bioreactor systems

Containment bioreactor systems designed for recombinant microorganisms require not only that a pure culture is maintained, but also that the culture be contained within the systems. Both GLSP and biosafety levels are detailed in this section.

A GLSP (Good Large-Scale Practice) level of physical containment is recommended for large-scale research of production involving viable, non-pathogenic and non-toxigenic recombinant strains derived from host organisms that have an extended history or safe large-scale use.

The GLSP level of physical containment is recommended for organisms such as those that have built-in environmental limitations that permit optimum growth in the large scale setting but limited survival without adverse consequences in the environment.

BL1-LS

A BL1-LS (Biosafety Level 1 - Large-Scale) level of physical containment is recommended for large-scale research or production of viable organisms containing recombinant DNA molecules that require BL1 containment at the laboratory scale.

BL2-LS

A BL2-LS (Biosafety Level 2 - Large-Scale) level of physical containment is required for large-scale research or production of viable organisms containing recombinant DNA molecules that require BL2 containment at the laboratory scale.

BL3-LS

A BL3-LS (Biosafety Level 3 - Large-Scale) level of physical containment is required for large-scale research or production of viable organisms containing recombinant DNA molecules that require BL3 containment at the laboratory scale.

No provisions are made at this time for large-scale research or production of viable organisms containing recombinant DNA molecules that require BL4 containment at the laboratory scale.

Steam Sterilisers

Autoclaves or steam sterilisers are used to sterilise items such as tools, fixtures and utensils used in aseptic processing. Modern systems are designed to fulfil the requirements of FDA and EU regulatory requirements. DIN 58950/58951 is a standard in which many manufacturers design and build steam sterilisers to fulfil the requirements set out in the document. Conformance to this standard ensures autocalves comply with the FDA and GMP directives. Industrial steam steriliser systems used in biotechnology companies comprise the following main components:

- Pressure container for sterilisation
- Vacuum pump
- PLC controller
- Human Machine Interface (HMI)
- Cycle software

The sterilisation process can be divided into three distinct stages:

> - **Pre-treatment Stage:** during this stage the autoclave begins to heat up and the air in the chamber is replaced by a mixture of steam and air.
> - **Sterilisation Stage:** the purpose of this stage is to kill any harmful microbes by using steam sterilisation. The temperature and pressure of the chamber is held at predefined settings for a specific period of time.
> - **After-treatment Stage:** cooling, decompression and drying occurs in this stage of the cycle.

The steriliser can be loaded with the help of a loading trolley manufactured with suitable materials or an automatic loading and unloading system. The steriliser can alternatively be equipped with trays for accommodating the goods to be sterilised.

Air Leakage Test (Vacuum) Test

The purpose of air leakage testing is to verify that the chamber is vacuum-tight and can maintain the vacuum over a period of time. To avoid loose interpretations, a formal definition of vacuum-tight should be documented. The British standard EN 285+A2 "Sterilisation. Steam sterilisers. Large sterilisers" provides definitions, guidance and a framework for testing steam sterilisers. The air leakage test should result in the chamber maintaining a predetermined pressure over a set period of time e.g. ten minutes.

Steam Penetration (Bowie Dick) Test

Steam penetration is tested using a Bowie Dick test kit. To verify the consistency of the process, this is typically done three times for a recipe or cycle.

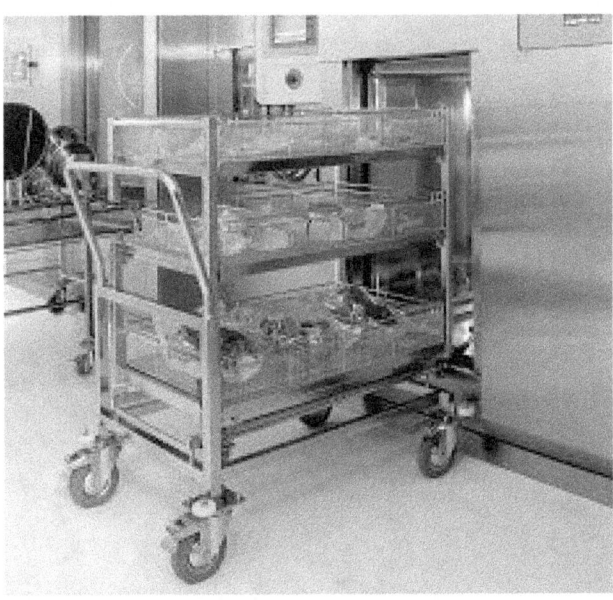

Figure 10: An autoclave trolley loaded with parts for sterilisation.

Pressure Leak Testing

This test is used to ensure the chamber does not leak. During the course of the test, the pressure is trended. The pressure drop over the test must be within specification. For example, the pressure decrease should be less than 100mbar during the course of the cycle (e.g. 10 minutes).

CHAPTER 4

DEPYROGENATION

What Is Depyrogenation?

Depyrogenation is a thermal process that involves the removal of pyrogens from components (e.g. vials or containers) that are used for injectable pharmaceuticals and biopharmaceuticals. A pyrogen is defined as any substance that can cause a fever. Bacterial pyrogens include endotoxins. Later on we shall see that endotoxins are used to challenge depyrogenation tunnels. Depyrogenation tunnel design varies depending on the manufacturer, however, they usually consist of the following components:

- Infeed and preheating
- Heating zone
- Cooling zone
- Outfeed and transport to next unit operation
- Automatic emptying

Pyrogens

Pyrogens are fever inducing proteins of low-molecular-weight proteins. Pyrogens of external origin are referred to as exogenous pyrogens. Modern injection and delivery systems are largely safe, yet adverse reactions are still reported. If a treatment or medication administrated via hypodermic needle is contaminated with toxins such as pyrogens fever can be induced which can lead to some death in some cases. It was known in the latter part of the 19th century that some parenteral solutions caused a marked rise in body temperature. The fever producing agents were not known, and hence described in general terms such as "injection fever," "distilled water fever," and "saline fever". Bacterial pyrogens are responsible for many of those early fevers and for many of the other biological effects described incidental to parenteral therapies.

The route of administration of a drug allows a pyrogen, if present, to bypass the bodies primary defences. The host's response is mediated through the leukocytes (white blood corpuscles) which in turn release their own kind of pyrogen (endogenous pyrogen) and this in turn initiates a fever like response and other biological reactions.

Bacterial Toxins

There are two general kinds of bacterial toxins: (1) endotoxins and (2) exotoxins. Endotoxins can be extracted from a wide variety of gram-negative bacteria. The term "endotoxin" is usually interchangeable with the term "pyrogen" although not all pyrogens. Higher doses of endotoxin are required to produce a lethal effect in the experimental animal than are required for exotoxins. The effects produced by endotoxins on the host are systemic such as fever and general body reactions, rather than strictly neurological effects, as is the case with most exotoxins. Endotoxins are found in the gram-negative bacteria mostly, and are obtained subsequent to the death and autolysis of the cells. The endotoxins are extracted from and associated with the cell structure (cell wall). Good examples of pyrogen producing bacteria are S. typhosa, E. coli, and Ps. aeruginosa.

Exotoxins are produced during the growth phase of certain kinds of bacteria and are liberated into the medium or tissue. Exotoxins are protein in nature and their reactions are specific. For example, Clostridium botulinum produces an exotoxin of unusual potency which affects only neurological tissue. Other well-known examples of exotoxins are tetanus toxin, Shiga toxin, and diphtheria toxin.

Properties of Pyrogens

Pyrogens are:

- Known to consist biochemically of a lipid-polysaccharide-peptide substance
- Heat stable at the temperature of boiling water
- Demonstrate a low order of immune response
- Produced from persistent gram-negative bacteraemia which could have a 50% mortality rate

Bactericidal procedures such as heating, filtration, or adsorption techniques do not eliminate pyrogens from parenteral solutions. All ingredients must be kept pyrogen-free in the first place. For this assurance, the manufacturer carries out comprehensive pyrogen screening tests on all parenteral drug ingredients and sees to their proper storage prior to use. Ideally, the manufacturer recognises the critical steps in the manufacturing operations that could allow growth of pyrogen producing bacteria and monitors these areas routinely. For example, the water in the holding tanks would be tested for pyrogens and the manufacturer would insist on minimum holding times so that only pyrogen-free water is used. Pyrogen-free water, as "water-for-injection" outlined in the USP, is the heart of the parenteral industry.

Pyrogen Assay - Limulus Amoebocyte Lysate

Many laboratories conduct pyrogen assays by means of the limulus amoebocyte lysate (LAL) test method. The LAL method is useful especially for screening products that are impractical to test by the rabbit method. Products best tested for endotoxins by LAL techniques are: radiopharmaceuticals, anaesthetics, and many biologicals. Essentially, the LAL method reacts hemolymph (blood) from a horseshoe crab (limulus polyphemus) with an endotoxin to form a gel. The quantity of endotoxin that gels is determined from dilution techniques comparing gel formation of a test sample to that of a reference pyrogen, or from spectrophotometric methods comparing the opacity of gel formation of a test sample to that opacity of a reference pyrogen. The LAL test is considered to be specific for the presence of endotoxins and is at least a hundred times more sensitive than the rabbit test. Even picogram quantities of endotoxins can be shown by the LAL method. Although LAL is a relatively new pyrogen testing method, it has produced a wide variety of polysaccharide derivatives that give positive limulus test results and also show fever activity. It is also a fact that some substances interfere with the LAL test even when pyrogens are present.

Some firms use the LAL test for screening pyrogens in raw materials, and follow up with pyrogen testing on the final product by means of the USP rabbit assay. The LAL test for pyrogens in drugs requires an amendment to the NDA on an individual product basis. LAL test reagents are licensed by the Bureau of Biologics. For devices, a firm must have its protocol approved by the Director Bureau of Medical Devices, before it can substitute the LAL assay for the rabbit. What is certain is that pyrogens remain a potential source of danger with the use of parenteral therapy.

Endotoxins and Depyrogenation

Endotoxins are used to challenge the effectiveness and consistency of depyrogenation tunnels. Endotoxin challenge vials must be processed through a depyrogenation process that must demonstrate a ≥3 log reduction in endotoxin. Typically, endotoxin challenge vials are placed in close proximity to thermocouples. Using this approach, the temperature profile of the position can be obtained during a cycle. Endotoxin challenge testing is often done during SAT and process development.

It is also a requirement of validation, however, no commercial product can be used during a depyrogenation tunnel performance qualification using endotoxins as the product would be potentially contaminated with endotoxins. Therefore, performance validation of depyrogenation processes results in the discarding of the vials or ampules. Depyrogenation tunnels generate tremendous amounts of heat and can operate up to temperatures as high as 320°C (depending on design and operational constraints). Air handling is also a key function of the depyrogenation tunnel. Tunnels should not allow non-sterile air from the room into the sterile air inside tunnel zones. This is done through the use of HEPA filters and an overpressure cascade approach of the tunnel compared to the surrounding room or environment. Air flow must be laminar in nature to ensure the tunnel can maintain the correct pressures and temperatures. Most tunnels are divided into two sections:

- Hot zone (depyrogenation)

- Cool zone (sterilisation/cooling)

The depyrogenation section typically operates at higher temperatures in excess of 270°C which is the recognised depyrogenation temperature. Depending on the technical specification of the components, set-points of 290°C, 300°C or 320°C can be used. Components move slowly through the depyrogenation stages of tunnels. The "sterilising cooling section" operation mode sterilises the cooling sections. Sterilisation cycle consists of the following steps:

1. Pressure drop
2. Draining heat exchanger
3. Heating up of cooling sections set value of temperature (e.g. 240°C)
4. Sterilisation cooling sections: keeping temperature at recipe set value for recipe set value of time
5. Cooling down without heat exchanger: until temperature reaches < 95°C
6. Cooling down with heat exchanger: until temperature reaches < 25°C

The key requirement of cool zone sterilisation is that the temperature within the zone is maintained at a minimum of 170°C for a period of no less than two hours. This gives a very high degree of assurance that the zone is sterile and suitable for sterile manufacturing operations to occur. In summary, the endotoxin challenge must be sufficient to demonstrate a ≥3 log reduction in endotoxin.

Biological Indicators for Dry Heat

Biological Indicators (BIs) (most commonly Bacillus atrophaeus) are used to demonstrate the efficacy of cool zone sterilisation in depyrogenation tunnels. Using a known indicator population and D-value, the delivered lethality needed to obtain an SAL of at least 10-6 can be determined.

The lethality of a cycle can be calculated using the below equation:

Lethality, $F(h) = \Delta T \times \Sigma L$

$L = 10(t-t_o) / Z$

$Z = 20$ constant

$t_o = 170$ the base temperature (°C)

t = actual temperature (°C)

ΣL = cumulative sum of time

ΔT = time differential (scan time)

Control of Materials

Items intended for sterilisation or depyrogenation should be prepared and maintained under conditions that will ensure that pre–sterilisation or depyrogenation levels of bioburden, particulate and pyrogen contamination are minimised.

Items that will come into contact with sterile dosage forms, filling equipment, containers and closures after sterilisation or depyrogenation in a dry heat oven should be packed for sterilisation in an appropriate clean environment. An appropriate standard would be environmental grade C or D under local protection by HEPA-filtered air.

Contamination Considerations

Protection of items against contamination before sterilisation or depyrogenation is not generally an issue when washers and tunnels are integrated. However, components should in all cases be received in packaging that minimises contamination risk (e.g., from fibres) and handled in such a way as to minimise contamination risk. Units should be loaded in compliance with the loading patterns established during cycle development. Load and date of sterilisation should be documented and identifiable.

Items should be clearly identified and controlled to avoid mix-ups between sterile and non-sterile items. Chemical indicators may be attached to containers or placed within loads.

These indicate, through a colour change, that items have been exposed to steriliser conditions but cannot be taken as proof of the adequacy of sterilisation cycles. However, if chemical indicators do not change colour they should be interpreted as confirming sterilisation failure.

Items that are not dried immediately after cleaning should be sterilised as soon as possible (no longer than eight hours and preferably within four hours of cleaning) to minimise the risk of microbial proliferation and eventually pyrogen formation between cleaning and sterilisation.

A maximum storage time before re-sterilisation should be specified in case the equipment is not used immediately.

Adequate cleaning, drying, and storage of equipment provide for control of bioburden and prevent contribution of endotoxin load.

Start-up Conditions

In sterilising ovens, following any drying phase, the load is typically heated up by closing the dampers to the fresh air supply. Air within the oven is continually recirculated over heating elements and through HEPA filters. In modern sterilisers the cycle is usually under automatic control. If the steriliser requires manual intervention for adjustments (e.g. dampers), then this should be very clearly and precisely defined in the operating SOP and details recorded on the record of each sterilisation cycle. In tunnels, the heating occurs as the components progress into the heating zone.

Control and monitoring should be independent, and operate from different temperature sensors. Normally, temperature control and routine monitoring is by fixed position chamber sensors. The relationship to load temperature is established by the validation. If there are movable permanently installed temperature sensors, then these should be placed within the oven chamber and within the most difficult to heat position of the load as determined during validation.

This should be very clearly and precisely defined in the operating SOP and details recorded on the record of each sterilisation cycle. Where only data from fixed position chamber sensors are available, the chamber sensor should be positioned in the same position as used in the validation, generally the most difficult to heat position of the chamber. An appropriate allowance for lag phase should be included in the standard cycle (e.g. to set the steriliser timer). This approach is used to compensate for load lag times (the time difference between chamber probes and the load cold point reaching sterilising temperature) as established during validation as part of performance qualification. Note that this correction for lag should be part of the standard cycle as defined by validation and incorporated into the operating SOP, included in the automatic or manual control (as applicable) and included as part of the master process record (mpr) or acceptance criteria used to assess the cycle.

Tunnel control and monitoring should be independent, and operate from different temperature sensors.
The control and routine monitoring of tunnels is by fixed position sensors. The relationship between tunnel and load temperature is established by the validation. The tunnel sensors should be positioned in the same position as used in the validation. The acceptance criteria for the cycle are set on the basis of the validation data such that the tunnel load receives the correct heat input. For tunnels, the sterilisation process is continuous and so the temperature record is of a set temperature. Thus, in order to verify heat treatment of components, the belt speed should be confirmed and, if adjustable, recorded either continuously or intermittently (at least at start and end during each day of operation).

In-Process Controls

Process parameters that are essential to sterility assurance should be verified and documented for every load processed. Other less critical process parameters that may be indicative of actual or potential steriliser failure should be verified at a lower frequency. Periodic checks:

- Confirmation of instrument calibration and performance of any applicable calibration checks
- Data to be obtained, documented and verified for each cycle
- Identification of the contents of the load
- Confirmation of compliance with validated loading pattern (ovens only)
- Confirmation of correct sealing of doors
- Confirmation of correct differential pressures
- Continuous record of the temperature, time, belt speed where applicable, throughout each cycle from at least one sensor

Cooling

Oven loads are generally cooled by switching off the heating elements and opening the fresh air dampers, which allows cool HEPA-filtered air to circulate around the load. The rate of cooling should be a compromise between rapidity and the need to avoid product damage. In particular, glass components may be adversely affected by internal stresses caused by rapid and uneven cooling. Note that the cooling phase is established in the validation and fixed as part of the standard cycle.

Failure of Depyrogenation

The checks on cycle records are vital as a failure of sterilisation or depyrogenation cycles may not be readily detectable in the product testing as there are no visible or practicable non-destructive means of testing for sterility.

The assurance of sterility is thus very heavily based upon the validated process conditions being consistently reproduced during routine operation. It is essential that any failures are promptly detected and that there is a clearly defined course of action in the appropriate operating procedure. Any cycle that does not meet any of its acceptance criteria should be thoroughly investigated.

Materials processed through such a cycle cannot be released solely on passing a test for sterility. Any abnormal or unusual occurrences should be formally recorded on the appropriate site documentation and notified to production management and quality management (even if the occurrence is not formally part of acceptance criteria). They should then be assessed for impact on the sterilisation or depyrogenation and on the functionality of the unit. Procedures should be in place to address such situations (e.g. containment measures). There must be a formal, thorough and fully documented investigation of all cycle failures under the site failure investigation procedure. Possible causes of sterilisation or depyrogenation failure(s) include but are not restricted to:

- Components held for insufficient time during sterilisation
- Too low of a depyrogenation temperature in the hot zone

This may happen in the event of the load lag time being longer than expected due to use of unapproved load patterns, over-loaded ovens, inadequate drying etc.

- Ingress of non-sterile air due to inadequate over-pressure, faulty door seals or filter failure
- Dampers failing to operate correctly
- Excessive residual water in containers (from washing stage)

CHAPTER 5

CLEANING AND DISINFECTION

Introduction

Cleaning and disinfection must take place according to defined procedures and programmes, with on-going environmental monitoring to ensure compliance to the microbiological limits and to detect the development of resistant strains of organisms. The effectiveness of all disinfectants must be validated with reference micro-organisms and local isolates. Hard surfaces of equipment, premises and materials that are decontaminated can be selected on a risk-based approach. The choice of disinfectants must be adequate to maintain good results on the viable environmental monitoring trend analysis.

What Is Meant by the Term "Cleaning"?

Cleaning can be defined as "the process of removing potential contaminants from equipment or product while maintaining the condition of equipment so that it remains fit for its intended use and is not subject to excessive damage due to the cleaning process." The words "grease and dirt free" are useful to remember as they give a practical understanding of what "clean" means. If equipment, parts or product become "dirty" they are often referred to as "soiled". Cleaning is a physical process where, particles, grease or organic matter is removed from a material or surface.

Why Clean Equipment or Products?

Equipment: Clean-in-Place, often abbreviated to CIP, allows equipment cleaning to occur with minimal disassembly of equipment. CIP programmes allow different products using similar or different materials to be manufactured on the same equipment.

Products: The supply of products and medical devices that are used by patients or healthcare professionals must be clean and free of contamination.

Regulatory Requirements: It is the aim of any manufacturer to provide safe and effective products for use by patients and end users such as doctors and nurses. Companies are granted licenses to supply markets with products based on regulatory compliance and product safety. Cleaning compliance is a key part of achieving a state of compliance and more importantly, supplying safe products.

Verification versus Validation?

Verification: Verification means confirmation by examination and provision of objective evidence that specified requirements have been fulfilled.[1] When it comes to cleaning, if the procedures have not been fully

---

[1] 21CFR820.3

validated, the effectiveness of the cleaning procedure should be verified at the completion of cleaning. This is "verification".

Validation: Validation means confirmation by examination and provision of objective evidence that the particular requirements for a specific intended use can be consistently fulfilled.[2]

Definitions

Clean Hold Time (CHT): The total time the parts are held clean post-cleaning.

Cleaning Agent: The chemical agent or solution used as an aid in the cleaning process.

Cleaning Process Parameters: The parameters that are critical in the cleaning process. Subsequent cleaning process monitoring may or may not utilise these parameters.

Critical Process Parameter (CPP): A control parameter that has a direct relationship to the quality, safety, effectiveness or performance of the intermediate or final product.

Dirty Hold Time (DHT): The total time the parts (or equipment) are held dirty prior to cleaning.

Maximum Allowable Carry Over (MACO): Amount of allowed product residue carry-over from lot-to-lot and batch-to-batch etc. This limit is based on the lowest of:
(1) limit based on toxicity, (2) limit based on the smallest therapeutic dose, and (3) worst case dose methodology.

Residue: Substance left on surfaces of equipment after cleaning that may pose a risk for subsequent use. Residues that may require cleaning include: product, excipients, raw materials/intermediates, non-volatile solvent, non-intrinsic cleaning agents such as detergents etc.

Worst Case Conditions: Considered to pose the greatest chance of process or product failure. The highest or lowest value of a given control parameter or set of parameters.

Visual Inspection: With regard to cleaning, visual inspection should be completed by appropriately trained and experienced personnel on completion of equipment/process clean down. Surfaces should be visibly clean and free of visible residue. Hard to clean places should be examined in particular.

Process Window: The selected operating range of machine settings/parameters that will produce product to meet all quality and product specifications.

Clean-in-Place (CIP): A cleaning method used to clean the inner surfaces of piping, vessels and process equipment without the need for disassembly.

PIC/S: The Pharmaceutical Inspection Convention and Pharmaceutical Inspection Co-operation Scheme (referred to as PIC/S) are two international bodies between countries and pharmaceutical inspection authorities, that co-operative in subjects relating to the field of GMP.

---

[2] 21CFR820.3

Skid: A modular process that can be plugged into a process onsite, with little construction or integration. Skids are used as part of clean-in-place solutions within the food and beverage and pharmaceutical industries.

Clean-in-Place (CIP)

Cleaning validation for a CIP system design involves the intersection of two similar or different products. For example, a pharmaceutical company manufactures two types of paracetamol caplets (tablets). Product A contains the active ingredient paracetamol, preservatives and other excipients. Product B is also a paracetamol product but it contains an additional ingredient, caffeine. Therefore, product B is branded differently and marketed with a more discerning customer in mind. Where multiple products are manufactured on the same equipment or machinery, the process is often referred to as non-dedicated. As with the above example, if the same equipment is used to produce product A and Product B, an intersection of products occurs.

Product A: Cleaning must be effective enough to remove residues to acceptable levels.

Product B: When manufacturing commences, the residue levels must not contaminate the product.

Residue is any substance or trace of substance left on equipment or surfaces after cleaning. It is near possible to remove all residue from surfaces so a residue limit should be medically safe and at a level that does not cause product quality issues or concerns.

Visibly Clean

Within any cGMP environment, the requirement to maintain a clean and suitable manufacturing area is key to compliance and ensuring product quality and customer safety. Visual inspection of the cleaning process must be done before swabbing. Inspection should confirm the equipment is visually clean and dry and no adverse odours are present. Upon completion of visual inspection, swabbing should then only be carried out if required by procedure. For areas that cannot be accessed for visual inspection or swabbing, a rinse sample can be taken in place of a swab. Sometimes it is not possible to obtain a swab or rinse sample, therefore visual inspection may be the only method used to verify cleaning effectiveness. In any validation an important theme is to challenge the consistency of a process. Samples must be representative to ensure a proper picture is painted. Sampling sites should be taken from "hard-to-clean" areas as well as "easy-to-clean" ones to ensure that samples are representative of the equipment.

Soils

"Soils" are a source of contamination to products and therefore can present a risk to patients or users. Soils can be introduced by unplanned and unintended events, but they are likely a part of the process or the result of a manufacturing agent being used within a manufacturing process. Examples would include coolant of cutting fluid used in a machining process.

The fluid is required to achieve a good surface finish and reduce tool wear. The presence of this soil on parts can potentially be:

- Dried on during the subsequent process step
- Compacted
- Dried on during dirty hold time
- "Baked" on during an oven process

With regards to CIP and cleaning between different products, cleaning should focus on product contact surfaces or process-critical indirect product contact surfaces. Non-critical cleaning of walls, floors and ceilings does not require the same level of cleaning. Likewise, dedicated equipment can often have a reduced cleaning programme.

Validation Strategies

In this section we examine validation strategies, a.k.a.

- Grouping approach
- Matrix/family approach
- Bracketing

A family or matrix approach can be used where similar products can be grouped together with a representative validation conducted to cover the particular grouping. With regard to medical devices, a particular product size or product configuration may be selected to represent the worst-case product. Therefore, by qualifying the worst case, all other products within the family of products would be considered validated. With regard to pharmaceutical products, e.g. solid dose manufacturing of pain killers, products of a similar chemistry/content can be grouped together. However, this approach must be clearly documented and technical rationale provided in advance of any qualification activities. This can be addressed in a validation plan or within a protocol. A grouping/matrix approach can be done by:

- Product (soil)
- Equipment
- Worst Case

Advantages of Grouping/Matrix Approaches

- Simplifies the amount of validation work
- Fewer validation runs

Typical target residues for CIP systems include:
- Drug (active)
- Cleaning agent

- Bioburden
- Endotoxin
- Degradation products or by-products

How Are Acceptance Levels Defined?

Several considerations need to be accounted for when establishing safe and effective residue levels including:

- The potential effects of target residue on subsequent products or raw materials
- Pharmacology of residue
- Toxicity of residue
- Stability issues

European guidance (Reference Human Drug CGMP Notes, 9:2, 2Q 2001) stipulates that equipment does not have to be as clean as the best possible method of residue detection or quantification, as absolute cleanliness is only required where feasible. However, it should be as clean as can reasonably be achieved –"to a residue limit that is medically safe and that causes no product quality concerns….".[3]

Uses of the Term "Limit"

L0= Daily amount allowed per patient (μg or mg)
(L zero)

L=1 Concentration in next product (mg/g)

L=2 Absolute amount in manufacturing vessel train (mg) [MAC – maximum allowable carryover] – L2

L3=Amount per surface area (mg/cm^2)

L4a = Amount per swab (mg)

L4b Conc. in swab extract solution (mg/g)

L4c = Conc. in "rinse" water (mg/g)

NOTE: L=0, Daily amount allowed is also known as:

Acceptable Daily Intake (ADI),

Acceptable Daily Exposure (ADE),

Permitted Daily Exposure (PDE)

---

[3] Human Drug CGMP Notes, 9:2, 2Q 2001

Safe Daily Intake (SDI)

Values for L0 can be a minimum daily dose of active 0.001, or a value based on toxicity data (MSDS sheets etc.).

PIC/S Guidance on Limits

<u>Who Are PIC/S?</u>

The Pharmaceutical Inspection Convention and Pharmaceutical Inspection Co-operation Scheme (jointly referred to as PIC/S) are two international instruments between countries and pharmaceutical inspection authorities, which provide together an active and constructive cooperation in the field of GMP.[4]

<u>What Is PIC/S Guidance?</u>

The most important point to remember when it comes to limits is that residues meet predefined criteria. The most stringent criteria are listed below:

(a) No more than 0.1% of the normal therapeutic dose of any product should appear in the maximum daily dose of the following (next) product.
(b) No more than 10 (parts per million, ppm) of any product will appear in another product (this value is not always the default).
(c) No quantity of residue should be visible on the equipment after cleaning procedures are completed. Spiking studies should determine the concentration at which most active ingredients are visible.[5]

---

[4] http://www.picscheme.org/
[5] http://www.picscheme.org/publication.php?id=4

MACO Calculations

There are two steps in calculating MACO residue levels. First, it is necessary to calculate the MACO from one batch to the next batch. The second step is to calculate the allowable "drug" or "residue" of each unique product contact surface for each piece of equipment. This then provides a calculation based on the overall equipment train (aka the equipment line).

The MACO is based on three calculations, which are:

(a) MACO based on toxicity
(b) MACO based on the smallest therapeutic drug dose
(c) MACO based on smallest solution batch size

(a) $$NOEL = \frac{LD50 \times NHW}{2000}$$
$$ADI = \frac{NOEL}{SF}$$
$$MACO = \frac{ADI \times SSBS}{LNDD}$$

NOEL = No Observed Limit Effect

LD50 = Lethal Dose of Drug

NHW = Nominal Human Weight

2000 = Is a constant factor for calculating NOEL

ADI = Allowable Daily Intake

SF = Safety Factor e.g. 1000

SSBS = Smallest Solution Batch Size

(b) MACO based on Smallest Therapeutic Drug Dose (STDD)

$$Product\ Carry\ Over = \frac{STDD}{SF}$$
$$MACO = Product\ Carry\ Over \times \frac{SSBS}{LNDD}$$

(C) MACO based on Smallest Solution Batch Size (SSBS)

$$Worst\ Case\ Number\ of\ Doses = \frac{SSBS}{LNDD}$$

$$MACO = LNDD \times Worst\ Case\ Number\ of\ \frac{Doses}{SF}$$

Using the above calculations, the MACO for the equipment train can be determined. The MACO for each individual piece of equipment of surface can then be calculated.

To calculate the MACO (allowable residue for each piece of equipment) you will need to have the following information available:

-MACO (per calculations above)

-Surface area of each piece of equipment and the total of the equipment train.

Example of calculation:

If a process has a MACO calculated to be 100μg and the total surface area of the equipment train is 100cm2

Surface Area 1 = 60cm2

Surface Area 2 = 40cm2

Surface Area 1 = 60% of MACO

Surface Area 2 = 40% of MACO

It is important to consider test methods and test method validation early on in the validation life cycle. A test method is a process or an action used to verify that a product feature or particular requirement meets a predefined specification. Test methods can be physical or analytical in nature. Test method validation should be completed in advance of cleaning as the test method is used to verify the outputs of such cleaning validations.

<u>Cleaning Process Design</u>

For Clean-in-Place (CIP), the key elements to be considered during the design stage include:

– equipment to be cleaned

– soils to be removed

– cleaning methods

– cleaning agents

– cleaning mechanisms

– cleaning parameters

– residue limits

Equipment Considerations

Firstly, for precision cleaning systems, the choice of equipment must be based on the intended purpose of the equipment; for example, will it be used for intermediate cleaning or for final cleaning? As you may expect, acceptance criteria for cleanliness are largely more stringent when it comes to final cleaning precision equipment. For CIP systems, the intended purpose of the equipment is a key design consideration. Materials of construction should be in keeping with the process, maintenance and cleaning regimes associated with it. Material certification should be provided by the supplier or vendor to ensure the correct grade of materials are used from approved suppliers.

In summary, the design should take into account:

- Difficult-to-clean locations
- What legacy systems are in place (hence knowledge)?
- Materials of construction
- Cleaning agents to be used
- Cleaning parameters
- Individual cleaning or cleaning as an equipment train

Critical Cleaning Parameters

The key parameters for cleaning can be remembered by using the acronym TACCT.

- **T**ime
- **A**ction
- **C**leaning chemistry
- **C**oncentration

➤ **T**emperature

Other cleaning parameters include flow rate, consideration of turbulence, water quality and rate of rinsing.

General Design Considerations

The efficacy of a cleaning process is influenced by the type of flow within the system; the two types of flow are laminar flow and turbulent flow.

Laminar flow is when fluid particles move in parallel layers, at a constant velocity.

Turbulent flow is when the movement of fluid particles varies in velocity and direction.

The Reynolds number of a system determines if the flow is turbulent or laminar. A Reynolds Number (Re) greater than 4000 is described as turbulent flow.

$$Re = \frac{316Q}{dK}$$

Q = volumetric flow, (gal/min)

d = internal diameter (inches)

K = viscosity (cP)

Dead Legs

A dead leg in the world of piping terminology refers to an area of piping where there is insufficient flow or a tendency towards water build-up or stagnation. The formal definition of a dead leg states that pipelines for the transmission of purified water for manufacturing or final rinse should not have an unused portion greater in length than 6 diameters (6D rule) of the unused portion of pipe measured from the axis of the pipe in use.[6]

Connections and Tie-ins

CIP systems must be connected to utility supplies such as process and de-ionised water. Precision cleanliness will require tie-in of water supply and drainage on a continuous or defined frequency. Welding is the preferred permanent method of connecting pipes. Non-permanent connections are also used which allow the disconnection and swap-out of piping, vessels and equipment. Orbital welding is a common method of welding when joining piping assemblies and vessels. Welding should meet necessary standards such as the visual weld criteria as detailed in the materials joining part of ASME BPE-2000 Standard. Discolouration of the weld can be evident as a result of the high degree of heat. The discolouration is a result of the oxidation and can result reduce the corrosion resistance of the weld. In general, welds should not exhibit cracks, crevices, or other surface deformities or visual defects.

Valves

Clamp-type connections can also be used for non-permanent connections. With regard to the use of valves, electromechanical valves that can be PLC controlled are preferred to manual valves. The level of automation depends on the complexity of the system. For example, a precision clean line used to clean metallic hip implants may have 5 or 6 clean and rinse tanks, all fitted with inlets and inlet valves. Having automated control is essential to running a complex line safely and efficiently.

Materials of Construction

When it comes to materials of construction, the same selection criteria can be applied to precision cleaning systems and CIP equipment trains. Above all, materials and their surfaces should be non-reactive, non-corrosive and non-porous. Stainless steel of a high grade is often the preferred material of construction. Examples of grades used include 304, 316 and 316L.

Cleaning Procedures and Methods

Validation of the cleaning procedures for the processing of equipment, including columns, should be carried out. This is especially critical for multi-product facilities. The manufacturer should have determined the degree of effectiveness of the cleaning procedure for each BDP or intermediate used in that particular piece of equipment.

---

[6] FDA guidance

Validation data should verify that the cleaning process will reduce the specific residues to an acceptable level. However, it may not be possible to remove absolutely every trace of material, even with a reasonable number of cleaning cycles. The permissible residue level, generally expressed in parts per million (ppm), should be justified and documented.

Cleaning should remove **endotoxins, bacteria, toxic elements, and contaminating proteins**, while not adversely affecting the performance of the column.

There should be a written equipment cleaning procedure that provides details of what should be done and the materials to be utilised. Some manufacturers list the specific solvent for each BDP and intermediate. For stationary vessels, often clean-in-place (CIP) apparatus may be encountered. After cleaning, there should be some routine testing to assure that the surface has been cleaned to the validated level.

One common method is the analysis of the final rinse water or solvent for the presence of the cleaning agents last used in that piece of equipment. There should always be direct determination of the residual substance.

Part of the answer to the question, "how clean is clean?", is, "how good is your analytical system?" The sensitivity of modern analytical apparatuses has lowered some detection thresholds below parts per million (ppm), down to parts per billion (ppb). The residue limits established for each piece of apparatus should be practical, achievable and verifiable. When reviewing these limits, ascertain the rationale for establishment at that level. Another factor to consider is the possible non-uniform distribution of the residue on a piece of equipment. The actual average residue concentration may be more than the level detected.

CHAPTER 6

PROCESS DEVELOPMENT

Vial Washers

The function of vial or ampule washers is to remove any debris or foreign matter prior to depyrogenation and further processing. This is achieved by exposing each glass vial or component to a series of internal and external rinses typically done with Purified Water (PUW) or water-for-injection. Removal of excess water is aided by compressed air blow outs or jets.

Developing a recipe that consistently and effectively ensures the entire surface area of the glass vial or ampule is the aim of process development. The vendor or manufacturer should provide baseline parameters and settings. However, there is often scope to improve key parameters and settings such as increasing pressures of flow rates. In most circumstances, performance qualification of vial washers are executed at worst-case settings. Therefore, process and cycle development studies should not only provide a production recipe but also a validation recipe consisting of worst-case conditions. While the "worst-case" settings provide an increased degree of challenge for the vial washer during validation studies, the settings should still deliver an effective coverage and washing of the components.

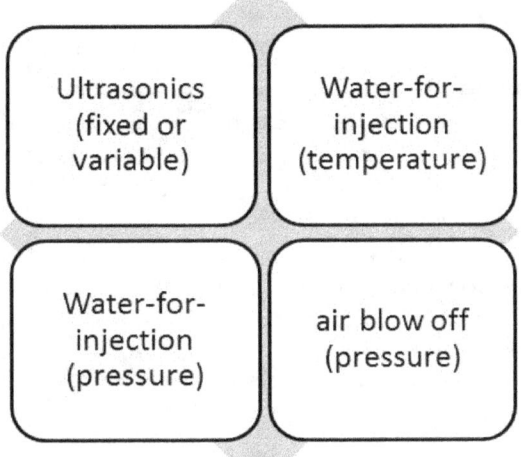

Figure 11: "Parameters of interest" for vial washers.

Depyrogenation Tunnels

In many instances, process and cycle development not only serves to develop and verify machine parameters, but also serves as a learning and knowledge gathering experience. Therefore, temperature mapping studies are valuable elements of any cycle development with regard to depyrogenation tunnels. In conducting temperature mapping, the temperature distribution profile of all sections of the tunnel can be learned. This can identify any potential "cold spots" or in general, provide confidence in the functionality of the equipment. Another aspect of depyrogenation tunnel cycle development is verifying the efficacy of the tunnel at its upper and low limits. In validation terms, this is referred to as Process-Operational Qualification (OQ-P), however, completing this during cycle development allows changes to be made more readily and free of the formal constraints of validation. Endotoxin test vials are used to show a reduction in levels as a result of depyrogenation. The following points should be addressed during cycle development:

- the minimum depyrogenation hold times can be reached
- the minimum temperatures can be attained (270°C)
- exit temperatures are suitable for the product being manufactured or filled
- temperatures within the tunnel are distributed evenly
- endotoxin challenge samples undergo a 3 log reduction
- confirmation of a positive pressure cascade

For cool zone sterilisation:

- a minimum temperature of 170°C can be reached and maintained for a minimum of 2 hours
- Fh, lethality values meet acceptance criteria
- biological indicators show no growth after a defined incubation period

Isolators

Critical to effective bio-decontamination cycles are the parameters and settings used in each stage of the cycle. As previously outlined, a bio-decontamination cycle has 3 stages; conditioning, bio-decontamination and aeration. During cycle development, the process parameters that result in successful bio-decontamination need to be verified. Through the use of temperature mapping, geometric mapping, empirical data and process knowledge, worst-case locations are selected and tested for VHP exposure (chemical indicators) and bio-decontamination (biological indicators). If a successful cycle is achieved, these parameters and settings can then be used during formal validation studies (PQs), production and commercial manufacturing.

The amount of positions (locations) must also be defined. Small isolators such as transfer isolators may only have one or two dozen test locations. Larger isolators containing processing stations such as filling/capping can have many more test locations.

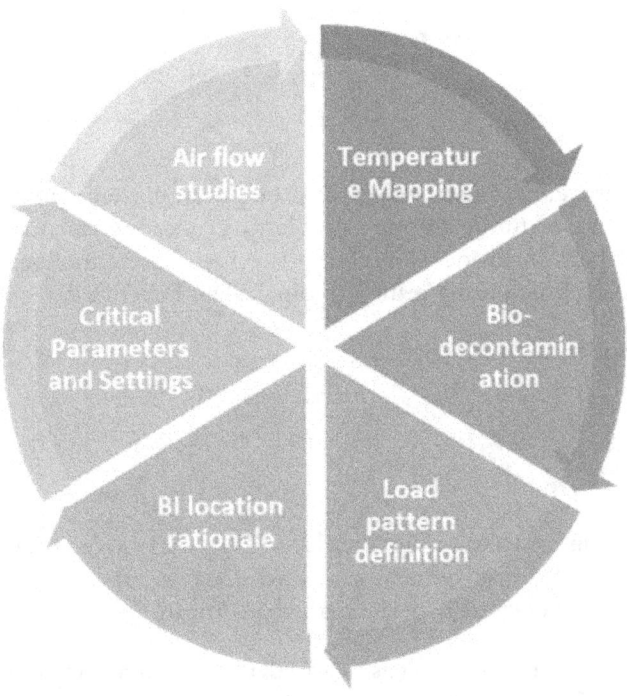

Figure 12: Key elements to cycle development and verification.

Critical Parameters

Critical process parameters are parameters than can negatively affect the outcome of bio-decontamination if not adequately defined. Isolators typically have a high degree of automation that controls and monitors air intake, dosing, temperature, air velocities and air pressures. The design and functionality of most isolators incorporate all of these conditions into the automated cycles or recipes.

Recommended Critical Process Parameters	Typical Units
Amount of H_2O_2 during conditioning	(g)
Dosing rate (conditioning)	(g/min)
Time for conditioning	(mins)
Amount of H_2O_2 during decontamination	(g)
Dosing rate decontamination	(g/min)
Time for decontamination	(mins)
Aeration time	(mins)

Rationale for Locations

For a new isolator system, procedures and recipes must be put in place to support manufacturing operations. Prior to validation or the manufacturing of any saleable products, the isolator must be verified to ensure it operates as intended. A key part of this development phase is to define the locations or positions that will be used for the placement of indicators during testing (biological and chemical).

Locations can be selected under the below headings:

-Critical positions
-Worst-case positions
-Geometric positions

Critical positions can include intervention points such as RTPs and glove surfaces. Any location where there is potential impact to product must be assessed and may merit the placement of BIs during bio-decontamination verification and eventually performance qualification.

Worst-case positions (temperature) can be based on data generated from temperature mapping studies. Temperature mapping using Kaye Validators can identify where the temperature may vary or fall outside the range of typical temperatures.

Worst-case positions (indicator) from studies with chemical or biological indicators can also identify positions where air flow or VHP may not penetrate or reach effectively. These present higher-risk areas that need to be assessed.

Geometric positions are positions that are selected to demonstrate the effectiveness across the area of the isolator. Air flow pattern analysis may also indicate high-risk areas.

Temperature Mapping

Temperature mapping is used throughout aseptic filling and manufacturing processes including autoclaves, ovens, depyrogenation tunnels and isolators. Using equipment such as Kaye Validators and thermocouples, a clear and accurate temperature map can be developed. For example, in order to temperature-map a thermal oven, a select number or thermocouples or probes would be placed at predetermined locations throughout the oven. The locations should represent the entire distribution of the oven and its boundaries. So essentially, thermocouples would be placed on the walls, doors and shelves of the oven. A cycle can then be started and the temperature throughout the oven over the course of the cycle can be collected.

Load Pattern Definition

Isolator load patterns are made up of tolls and materials that must be present within the isolator during manufacturing. Examples are listed below. Multiples of the same item are common as the restricted nature of isolators requires positioning of such items in key positions to allow access through glove positions and so on. Some common tools and materials:

- Slider tool
- Forceps (multiple sizes)
- Scissors
- Settle plates
- Contact plates
- IPA wipes
- Lint-free dry wipes
- Sterile markers
- Sterile tape
- Ziploc bags

Prior to the operation of an isolator or aseptic processing, a load pattern must be defined and documented. The load pattern is made up of tools and utensils that are needed during normal operation and production. Common items include forceps, scissors, settle plates, zip-lock bags, sterile IPA wipes. All items must be wiped with IPA prior to placement within the isolator enclosure. Once defined and documented, the VHP cycle (bio-decontamination) must be validated for the isolator load pattern.

Points to note:

-Isolator should be clean and free of debris or rejects.

-Materials and tools must be cleaned with IPA.

-The materials of construction (MOC) should be documented and compatible with VHP.

-The load pattern should be clearly defined in a standard operating procedure or work instruction.

-Glove integrity is required in order to allow safe operation of the isolator.

-The gloves are pre-tested to ensure the integrity of the gloves to allow safe and effective operation.

Using VHP (Vapourised Hydrogen Peroxide), a bio-decontamination cycle is completed on the isolator to decontaminate the isolator and the contents of it (utensils, filling stations, capping stations etc.) Once a successful VHP cycle is completed, the isolator and filling line are considered sterile and will remain sterile for the duration of the campaign. Autoclaved parts can be transferred aseptically in to the isolator during a campaign using rapid transfer ports (RTPS). The RTPs can be used to aseptically remove parts from the isolator also.

Alarms and Alarm Handling

Glove port intrusion should result in appropriate light curtain barriers triggering an alarm.

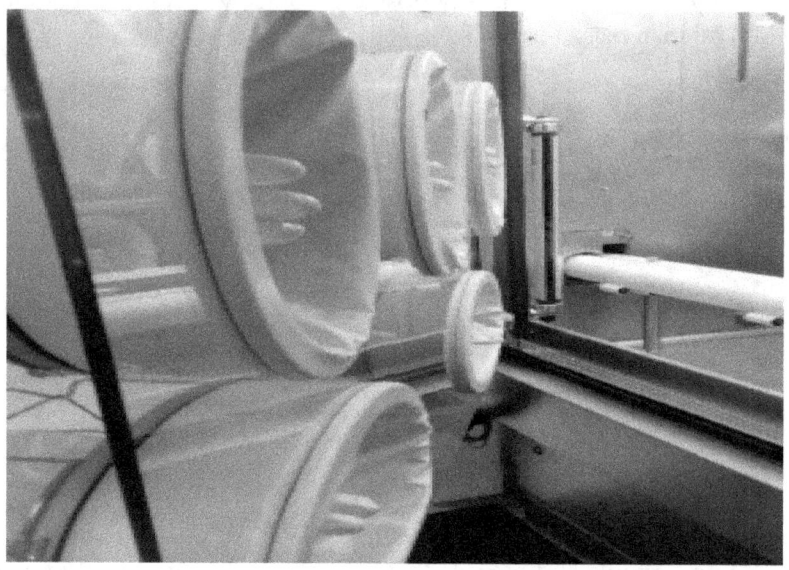

Figure 13: Interior side of an isolator door.

Smoke Studies on Depyrogenation Tunnels

Depyrogenation tunnels are logically positioned upstream to filling lines and container closure unit operations. As previously described, depyrogenation tunnels process glass vials, ampules or cartridges to ensure they are sterile and endotoxin free.

A design requirement of tunnels is that they can create a pressure differential inside and this can be maintained during normal operating conditions. The positive pressure differential is designed to ensure non-sterile air does not enter the sterile cool zone of the depyrogenation tunnel. Smoke studies are performed to ensure proper airflow direction. The differential air pressure in the hot zone should always be positive relative to the infeed zone, cool down zone, and to the room or surrounding environment. Some studies also serve to demonstrate that there is no air ingress into the sterile cooling zone from outside the tunnel.

CHAPTER 7

PHYSICAL PROCESSES

Fluid Flow

Fluids and fluid flow play a key role in bioprocesses since most of the required physical, chemical and biological transformations take place in a fluid phase. A fluid is a substance that undergoes continuous deformation when subjected to a shearing force. For example, a shearing force applied to stacked plates will cause them to move over one another.

A Shear Force Must Be Applied to Produce Fluid Flow

Classification of Fluids

Fluids can be either (1) gases or (2) liquids. Two physical properties are used to classify fluids; these are:

- Viscosity of the fluid
- Density of the fluid

If the density changes with pressure, the fluid is compressible. Gases are generally classed as compressible. Liquids are generally classed as incompressible. Viscosity is the most important property affecting the flow characteristics and behaviour of a fluid. The viscosity of fermentation fluids is affected by the presence of cells and gases. Simply put, viscosity is the resistance to flow.

Mixing

Mixing is a key part of processing in bioprocessing. For example, in fermentation, mixing allows cells to come into contact with the dissolved nutrients and oxygen. It also contributes to controlling the culture temperature, ensuring the temperature is distributed throughout the mix. Some negative consequences of mixing include cell damage as a result of the force. This can result in a loss in productivity and is often a challenge between pilot operations and scale-up to commercial manufacturing. Some key applications of mixing include the following:

- Blending
- Dispersion of gases such as air through liquids in the form of small bubbles
- Maintaining suspension of solid particles such as cells

> Promoting heat transfer to or from liquids

Mixing is most likely to be conducted in cylindrical stirred tanks. Most tanks include baffles which are strips of metal mounted against the wall of the vessel. Mixing is done via an impeller mounted on a centrally located shaft.

Vessel Geometry

The shape of the base of stirred tanks is an important design feature. It has a direct impact on the efficiency of mixing. It should be rounded at the edges to prevent areas where the currents cannot reach. These areas are often termed stagnation zones. To allow for efficient mixing with one impeller of Diameter Di, equal to a ¼ to ½ the tank diameter (Dt), the height of the liquid in the tank should be no more than 1.0 to 1.25 Dt.

Another aspect of vessel design and geometry that impacts mixing efficiency is the clearance between the impeller and the lowest point of the tank (floor). In practice, the clearance, Ci, is within the range of 1/6 to 1/2 the tank's diameter.

Baffles

Baffles are common features within stirred tanks. They help the mixing of the tank's contents and create turbulence by breaking up circular flow which is a result of the impeller rotation.

Sparger

The purpose of the sparger is to allow the controlled release of small bubbles into the tank. A simple sparger can be a perforated tube, or a single orifice sparger. As a rule of thumb, as the size of the vessel increases, the gas requirement also increases. As bubbles are released and rise up from the outlet, they move into the impeller zone. Within this zone, the bubbles break up and disperse within the tank.

Stirrer Shaft

The purpose of the stirrer shaft is to rotate the impeller. This is achieved by the shaft transferring torque through the shaft to the impeller. The shaft must be designed to adequately hold the stirrer's weight and also resist any deflection. Incorrect specification of the shaft can lead to inefficient designs and catastrophic failure. It is also important that the shaft operates with excessive vibration.

Heat Transfer

For bioprocessing, heat exchange generally occurs between fluids. Heat exchangers allow the transfer of heat through a solid metal wall which acts to separate the fluid or product. The important factors that facilitate proper heat transfer include:

- Design consideration of the surface area
- Agitation and turbulent flow of fluid
- Temperature range

There are different heat transfer solutions available to the bioprocess engineer, however, the most common is the double-pipe heat exchanger. The double-pipe heat exchanger uses two metal pipes, one inside the other. When one of the fluids is hotter than the other, the heat is exchanged across the wall.

Cocurrent Flow: This is when the fluids are applied in the same direction.

Countercurrent Flow: This is when the fluids are applied in different directions.

CHAPTER 8

EQUIPMENT VALIDATION

Introduction

Validation is a legal and regulatory requirement for the manufacture of medicinal products. The area of validation can be sub-divided into two elements. Equipment qualification (eq) and process validation. Equipment qualification ensures that equipment operates as intended and is installed in accordance with the manufacturer's recommendation. Process validation involves the provision of documented evidence to confirm a particular process performs consistently and meets predetermined specifications.

All equipment that can impact the quality of product is subject to validation, hence equipment and systems used in aseptic manufacturing must undergo equipment and process validation.

Installation Qualification (IQ) protocols should cover verification that all utilities are installed correctly to the manufacturer's recommendations. All sitting and mechanical connections should also be confirmed as adequate. Other key tests and verifications include:

- Documentation of Materials of Construction (MOC)
- Calibration of equipment based instrumentation
- Spare parts listing
- Preventative maintenance schedule creation
- Electrical installation verification
- Health and Safety assessment
- Ergonomic assessment
- Documenting software and hardware
- Backup of software
- Backup of recipes (sterilisation, bio-decontamination etc.)

The system User Requirements Specification (URS) should provide the basis of testing and must be fulfilled during the course of validation.

Materials of Construction (MOC)

The materials of construction and evidence of the same (certificates) form part of installation also. Materials must be fit for the intended purpose and compatible with products and manufacturing agents that come into contact with them.

For example, fermenters are made of materials that are suited to the use of steam sterilisation techniques and regular cleaning. Such materials can be classed as both non-reactive and non-absorptive surfaces. Most aseptic processing equipment that incorporates product contact surfaces is made of high-grade stainless steel. Cheaper classifications of stainless steel can be used for jacketing and other non-product contact areas.

All interior product contact surfaces should be polished to a "mirror" finish. Welds also need to be finished in a similar manner. Electro polishing provides a better quality surface finish than mechanical polishing.

As with any chemical reaction, factors such as temperature, pH and oxygen concentration can impact the performance and yield. To ensure the optimum conditions are maintained, it is important to monitor and control such parameters and factors. By far the most common these days is automatic control of systems and equipment with automatic feedback and adjustment.

Operational Qualification (OQ) is the second component of equipment qualification. This is "establishing by documented evidence that the equipment operates per specifications and over the required ranges and to required tolerances". Equipment is also tested to ensure alarms and controls operate as required and intended. Some typical checks included in an equipment-operational qualification are testing of alarms, control system testing, utility failures and functional and operational testing.

Suggested IQ/OQ Verifications/Tests

Standing Operating Procedures (SOPs)

SOPs are designed to provide formal documented instruction on how to execute tasks or operate equipment or machinery. While each company will require different headings, a work instruction or SOP should cover set-up, system operation, cleaning and shutdown to name but a few.

Test Instrumentation Calibration

External test devices such as temperature probes, volt meters, lux meters and particle counters may be required to take measurements during an equipment qualification. Test instrumentation should have a suitable range, resolution and accuracy. Certificates of calibration should also be available, with calibration conforming to a recognised external standard. Information such as the serial number, model number and manufacturer should be recorded for reference and traceability.

Equipment Based Instrumentation Calibration

All equipment based instruments must be calibrated as part of equipment validation or in advance of it. Instruments should have a unique calibration ID.

Electrical Checks

Appropriate connections and earthing checks are essential. A review of electrical drawings to ensure the physical status is as per drawings and specifications is also required. Cables and electrical hazards should also be appropriately labelled.

Mechanical Checks

Ensure the systems are fixed, fastened and integrated mechanically. Safety guards and barriers should also be in place where required.

Pneumatic Checks

Verify the proper supply and integration of compressed air. Supply should be leak-free, regulated with filters and water traps fitted as required.

Documentation

Verification that design, operation and maintenance documentation has been received from the manufacturer and is stored appropriately.

Ergonomics

Controls and HMIs should be positioned to facilitate ease of use and should be identified clearly.

Health and Safety

All hazards are identified and guarded, pin points are identified. No trip hazards are evident and emergency stops function.

Software

Equipment installation software checks should record the names and version numbers of all software. HMI software, PLC software, application software. Provision should be made for disaster recovery and backup.

Hardware

Computer hardware should be recorded to include the model, manufacturer, serial number and specification details.

Environmental

Any features detailed in the URS relating to environmental requirements need to be verified during IQ/OQ. For example, automatic shutdown after periods of inactivity. Heating and cooling systems should also be appropriately insulated.

Alarms

Automated processes such as sterilisation tunnels, autoclaves, filling machines and isolators typically have many alarms and controls. Alarms can be categorised as critical or non-critical to the process or product. Depending on the vendor or manufacturer, alarms can also be grouped according to the type of alarm (EHS, process, mechanical, pneumatic and so on). Alarms should be tested to ensure the right action by the machine is taken, the process comes to a safe stop, and that the alarm can be acknowledged and the alarm condition cleared.

<u>Utility Failure</u>

Also referred to as provoke testing, utility failure of compressed air, fume extraction, electrical supply and so on is to ensure that in the event of failure during commercial manufacturing, the equipment comes to a safe stop and can be brought back into use upon recovery of the utility.

<u>Fixtures</u>

Materials of construction must be suitable for the intended use. In aseptic processing, high grades of stainless steel (316L) are the preferred material of use.

<u>Functional Tests</u>

The individual functions of equipment must be verified during commissioning and qualification.

Depyrogenation Tunnels (Equipment Validation)

While many of the general installation and operational checks will be relevant to depyrogenation tunnels, a number of key focus areas must be verified during equipment IQ/OQ of depyrogenation of tunnels. Temperature: IQ/OQ tests should verify the capability of the tunnel to reach and maintain set temperatures over the operational range. Overshooting or undershooting of set temperatures should not be observed. The temperature should be consistent over the operational area of the tunnel in which product travels. Alarms must also be verified to ensure upper and lower alarm tolerances are responsive.

Pressure: For equipment qualification, pressure transducers and gauges must be calibrated and functionally verified. Maintaining positive pressure is a key element of depyrogenation design. Other equipment IQ/OQ tests for depyrogenation tunnels include:

- HEPA filter integrity
- Pressure differential control

Isolators (Equipment Validation)

In addition to the standard IQ/OQ equipment checks and verifications, the following list details isolator specific verifications that should be considered:

- Air-flow pattern, empty, in normal operation and in bio-decontamination mode
- Air-flow pattern in empty and decontamination mode
- HEPA filter integrity
- Leak detection
- Leak test of isolator

- For isolator units configured with transfer isolators, maintenance of pressure differential that ensures the pressure in the workstation remains positive when passing items between the workstation and the transfer isolator
- Confirmation of the injection rate with decontamination gas concentration
- Verification of the stage of recipe or cycle
- Temperature mapping
- Gas generation test

Steam Sterilisers (Equipment Qualification)

Approved user requirements specifications, regulatory requirements and technical standards contribute to the scope and content of equipment validation. For autoclaves, the functionality of cycles must be verified to ensure the correct system sequence. Critical operational verifications include:

- Verification cycle attains a temperature of 121 °C
- Temperature distribution tests
- Air filter temperature mapping
- Small load thermometric tests

HEPA Filters

HEPA filters are composed of a mat of randomly arranged fibres. The fibres are typically composed of fibreglass and possess diameters between 0.5 and 2.0 micrometres. Key factors affecting its functions are fibre diameter, filter thickness and face velocity. The air space between HEPA filter fibres is typically much greater than 0.3 μm.

The common assumption that a HEPA filter acts like a sieve where particles smaller than the largest opening can pass through is incorrect and impractical. Unlike membrane filters at this pore size, where particles as wide as the largest opening or distance between fibres cannot pass in between them at all, HEPA filters are designed to target much smaller pollutants and particles. These particles are trapped (they stick to a fibre) through a combination of the following three mechanisms:

Interception

Where particles following a line of flow in the air stream come within one radius of a fibre and adhere to it.

Impaction

This is where larger particles are unable to avoid fibres by following the curving contours of the air stream and are forced to embed in one of them directly; this effect increases with diminishing fibre separation and higher air flow velocity.

Diffusion

Diffusion is an enhancing mechanism that is a result of the collision with gas molecules by the smallest particles, especially those below 0.1 μm in diameter, which are thereby impeded and delayed in their path through the filter; this behaviour is similar to Brownian motion and raises the probability that a particle will be stopped by either of the two mechanisms above; this mechanism becomes dominant at lower air flow velocities.

Diffusion predominates below the 0.1 µm diameter particle size. Impaction and interception predominate above 0.4 µm.

In between, near the most penetrating particle size (MPPS) 0.21 µm, both diffusion and interception are comparatively inefficient. Because this is the weakest point in the filter's performance, the HEPA specifications use the retention of particles near this size (0.3 µm) to classify the filter.

However, it is possible for particles smaller than the MPPS to not have filtering efficiency greater than that of the MPPS. This is due to the fact that these particles can act as nucleation sites for mostly condensation and form particles near the MPPS.

Lastly, it is important to note that HEPA filters are designed to arrest very fine particles effectively, but they do not filter out gasses and odour molecules.

Today, a HEPA filter rating is applicable to any highly efficient air filter that can attain the same filter efficiency performance standards as a minimum and is equivalent to the more recent NIOSH N100 rating for respirator filters. The United States Department of Energy (DOE) has specific requirements for HEPA filters in DOE regulated applications. In addition, companies have begun using a marketing term known as "True HEPA" to give consumers assurance that their air filters are indeed certified to meet the HEPA standard.

CHAPTER 9

PERFORMANCE QUALIFICATION

Depyrogenation — Performance Qualification (PQ)

Performance qualification examines the effectiveness and reproducibility of the depyrogenation cycles in respect of a particular load. Thermocouples are used to verify the set temperatures are reached and maintained throughout the cycle and at various points in the tunnel to demonstrate consistent heat distribution. Endotoxin challenge vials or ampules are also used to demonstrate reduction in endotoxin levels to within acceptable levels and hence ensuring products are safe for patient use. Traditionally, the theme of consistency in performance qualifications has been demonstrated by three completed distinct batches or runs as part of PQ. Although risk-based approaches to validation (e.g. ASTM E2500 etc.) are increasingly used across the life science industry, completing a minimum of three batches or runs for initial performance qualification is still the expected requirement.

The size of the glass components (vials, ampules or cartridges) must also be considered in the design of performance qualifications. The efficacy of the tunnel or sterilising effect is ultimately determined by the size, shape and mass of the components processed. A family or bracketing approach may be utilised to reduce the amount of runs or cycles required. For example, a manufacturing process utilising four different vial sizes across a product – 5ml, 10ml, 15ml and 20ml. Based on technical rationale and some level of evidence, the 5ml and 20ml vials could be considered "worst case". The 5ml being the smallest may exhibit the smallest nick sizes and internal geometry. The 20ml would in this case be the largest vial, with the largest surface area and mass – another worst case configuration.

Another consideration is the position and quantity of thermocouples. For PQ, a rationale based on sound science must support the locations and quantity of thermocouples used during the validation. This should be based on historic data if available, or in the case of new equipment, data generated during SAT testing and/or engineering development studies.

The thermocouple placement should also be carefully considered. If a particular location is to be assessed, the thermocouple should be secured with Kapton tape. Movement of the probe during a test cycle may result in the data being inconclusive or deemed non-representative.

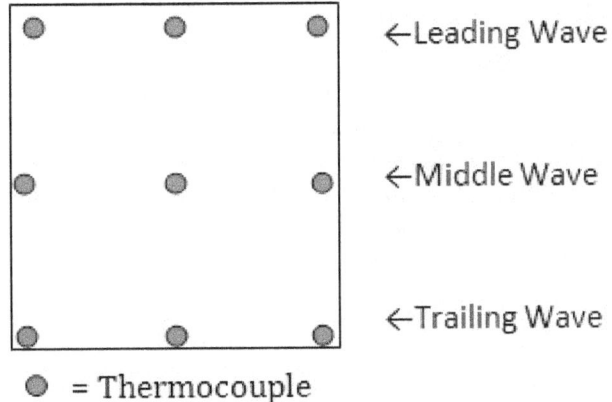

Figure 14: Simple representation of a depyrogenation tunnel (rectangle). Tunnels are filled with components which progress along a belt. Therefore, each batch shall have a leading wave, middle wave and trailing wave. Place thermocouples at the above locations should verify the temperatures at each location as the product moves through the depyrogenation tunnel. Endotoxin challenge samples may also be placed in similar locations.

Thermocouples – Points to Consider

- The point of contact with the component (e.g. bottom of vial, middle of vial etc.)
- Size of component if a bracketing approach is chosen
- Are there cold spots within the tunnel
- Rating of thermocouples (depyrogenation temperatures can well exceed 290°C)
- Length of thermocouples should allow for the travel through the length of the depyrogenation tunnel
- Number of locations probed (large depyrogenation tunnels require increased data points)

Cool Zone- Performance Qualification

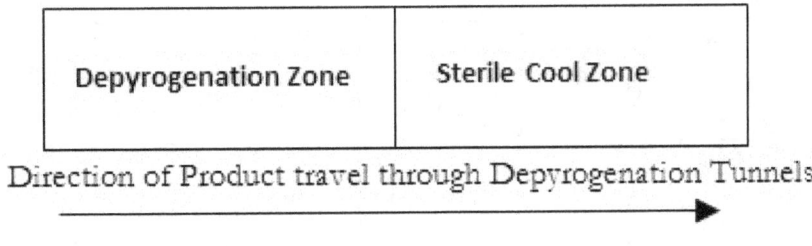

Figure 15

As illustrated above, performance qualification of depyrogenation tunnels can be divided into two elements (1) depyrogenation and (2) cool zone sterilisation. After glass vials or ampules are processed through the depyrogenation zone, they then enter the "cool zone" where temperatures are typically lower. However, prior to the transfer of product into the cool zone, a sterilisation cycle is typically performed. This consists of setting the cool zone temperature to a known temperature for a defined period of time. For example, a cool zone sterilisation cycle may require at set temperature of 230°C for two hours.

For initial performance qualifications, three consecutive runs are typically completed for newly installed equipment. Cool zone PQs are completed using temperature mapping equipment such as Kaye Validators. Specific locations are probed with temperature probes and a cool zone sterilisation cycle is run. The positions of thermocouples should be based on technical rationale developed in earlier process and cycle development. The positions must be representative of the size and shape of the tunnel. For example, positions that may merit the placement of heat penetration thermocouples may include:

- Tunnel belt
- Interior panels /walls
- Doors (separation depyrogenation zone from cool zone)
- Extremities of the cool zone

BIs can also be placed in positions identified as "cold spots". If the sterilisation cycle is effective in rendering biological indicators located at the coolest positions sterile, then it can be concluded that positions showing a higher temperature are also found to be sterile.

Recommended Depyrogenation Performance Qualification Tests:

Parameter/Description	Recommended Acceptance Criteria
Depyrogenation Hold-time: defined as the total time in which the tunnel reaches and maintains a set temperature during a cycle, where the temperature of vials is held at a minimum of 270°C	≥102 seconds at 270°C
Endotoxin Challenge: a reduction in levels of Endotoxin (of Endotoxin spiked samples)	≥ 3 log reduction
Exit temperature: the temperature of vials/ampules upon exiting the coolzone into the infeed of the isolator.	≤25°C or as required based on the product. Some proteins and treatments are sensitive to moderate temperatures

Recommended Cool Zone Sterilisation Performance Qualification Tests:

Parameter/Description	Recommended Acceptance Criteria
Dry Heat Hold-time: the total time in which all locations in the tunnel are ≥ 170°C.	≥120mins at 170°C
Fh Calculation: a measure of sterilisation time	≥ 32 mins

F Value Definition

A term used to describe the delivered lethality calculated based on the physical parameters of the cycle (e.g. F0, FH). The F-value is the integration of the lethal rate (L) over time: The lethal rate is calculated for a reference temperature (T ref) and z-value using the equation $L=10(T T ref)/Z$.

Lethality Definition

The capability of the sterilisation process to destroy micro-organisms, i.e. achieve a 106 reduction in population of Bacillus Stearothermophilus.

☐

Isolators

With reference to PIC/S guidance (PI 014-3, Isolators used for Aseptic Processing and Sterility Testing) placing three or more BIs at defined worst-case locations is recommended for performance qualification. Depending on the size of the isolator, BI locations may range from 20 to 90 different positions or more.

If there are any positive growths the proportion of positive to negative can be used to estimate the number of survivors and thus calculate log reductions. Given this information, any variation in the process is estimated and the significance of it can be evaluated. If there is only one BI in each position, and only growth/no growth is established, then the number of any survivors is unknown and the size of the possible variation in the process cannot be estimated. The BI challenges demonstrate that the VHP cycle parameters to be qualified are sufficient to ensure a six log reduction in a known microbiological system.

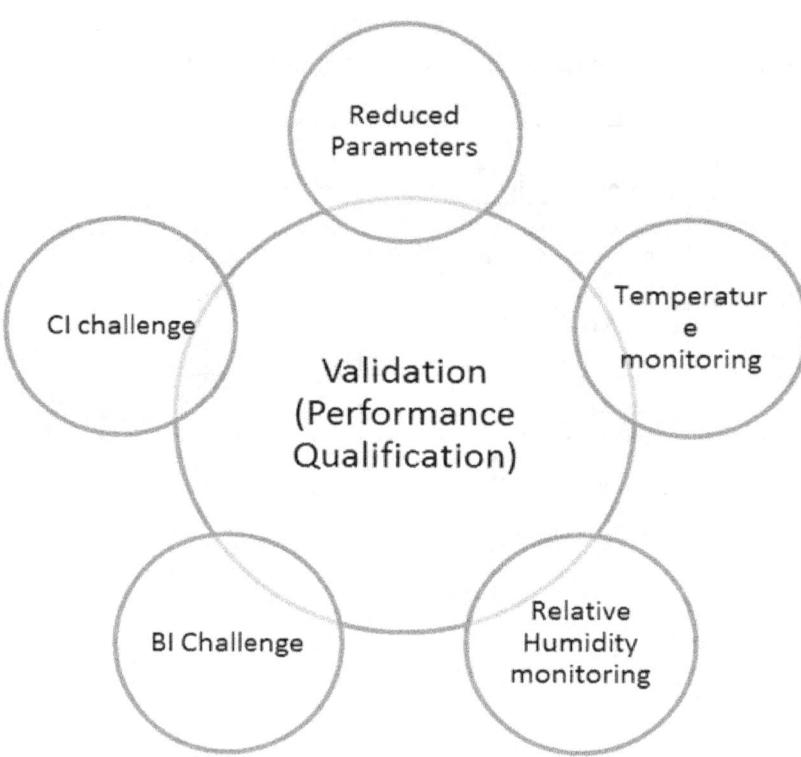

Figure 16

- BI challenge is achieved by placing biological indicators within the barrier system.
- Chemical indicators are included. These change colour when exposed to the decontamination agent.
- Temperature readings are captured with data loggers.
- Relative humidity readings are captured with data loggers.
- Reduced process parameters are controlled via a recipe.

Reduced Parameters

As an output of any engineering development (aka cycle development) a production recipe or cycle should be documented. One would have proven the efficacy of the recipe or cycle during the testing or runs completed.

However, regulations require manufacturing processes to be validated. Once the prerequisites to validation are completed (equipment IQ/OQ and engineering/cycle development) performance of the isolator barrier system is required. Execution of performance qualification at worst-case parameters is achieved by reducing one or more critical parameters to less than normal production parameters. For example, reducing the amount of VHP dosing during the bio-decontamination stage of a cycle would result in a greater challenge. (Less VHP may lead to a positive BI result or no change in chemical indicators.) If a production recipe has a dosing time of 20 minutes, reducing the dosing time to 15 minutes during validation runs would constitute a "worst case".

Isolator Load Verification

To ensure that the load pattern qualified under performance qualification is accurate and according to established procedures (from engineering and cycle development) load verification should be completed. This simply means the items that make up the load are placed in their correct positions within the isolator. Visual confirmation ensures that the right tools and materials are in the right positions, at the right quantities. Load components must be at the exact locations as detailed in SOPs or the relevant validation protocol (e.g. performance qualification protocol).

Isolator VHP Biological Indicator Challenge

VHP BI challenges are used to demonstrate that the VHP cycle parameters are capable of achieving a six log reduction from a known microbial population.

In order to prove consistency, BIs (Biological Indicators) are often placed in triplicate at predefined positions with the isolator. Once again, these positions should be based on scientific rationale and backed by documented evidence from either engineering runs or process development studies

After a bio-decontamination cycle has finished, test BIs must be removed from the isolator. As the BIs should now be rendered "sterilised", care should be taken in order to prevent contamination through poor handling practices.

- Wear sterile gloves and sanitise with 70% IPA
- Allow IPA to dry prior to removing BIs from BI location
- Use aseptic technique to transfer BIs into sterile centrifugal tubes
- Sanitise hands with 70% IPA after BI is removed and prior to handling the next BI

Chemical Indicator Challenge

The chemical indicator study will be performed for each run to verify uniform exposure of VHP through the filling and closing machine isolator.

Chemical indicators are placed throughout the filling and closing machine isolator with each group of triplicate BIs. The VHP cycle will be run and once the run is complete the CIs will be removed and inspected.

Chemical indicators yield qualitative information by changing colour. The indicators generate qualitative information enabling verification that the decontamination agent has reached all the points to be decontaminated.

All the chemical indicators shall display colour change at the end of the decontamination phase to indicate exposure to VHP.

Temperature and Humidity Study

The purpose of temperature and humidity mapping is to record the temperatures and humidity within an isolator during the conditioning and bio-decontamination phases of the cycle. Both temperature and humidity can influence the effectiveness of a the cycle. A well designed isolator will ensure a relatively stable temperature and humidity in particular at the pre-conditioning stage. The outer environment in which the isolator sits can also impact both temperature and humidity.

During process development, Kaye Validators and thermocouples may have been used to record temperatures. However, during validation of isolators, data loggers are more suitable. Data loggers can be placed in predetermined positions to capture both temperature and humidity.

Prerequisites to Execution of a Bio-Decontamination Run:

- Record the lot number and item number of the VHP sanitant
- Record the expiry date of the VHP
- Confirm isolator has been sanitised using IPA 70/30
- Record the lot number and item number of the IPA 70/30
- No debris

H2O2 Monitoring

Vapourised hydrogen peroxide (H2O2) used during bio-decontamination has the potential to cause harm to personnel and operators working in the vicinity of isolators. Ambient H2O2 levels are monitored using extremely responsive sensors that pick up minute traces of VHP in the atmosphere (to within 2-10 part per million (ppm)).

If a VHP sensor picks up levels of VHP within its working envelope, it typically triggers an alarm. The isolator system should also come to a safe stop. For example, if the cycle is in the bio-decontamination stage where VHP is being delivered to the isolator, the system should immediately stop this process and safely abort the cycle.

Internal H2O2 sensors also allow ongoing monitoring of concentrations during a cycle. Alert levels and action limits are typically defined in a cycle recipe which is used to define the duration and parameter requirements.

Figure 17: Gas detector that monitors the oxygen levels within a room.

Critical Parameters Monitoring

Critical parameters such as humidity, pressure, temperature and air velocity should be documented in an approved and controlled SOP. It is also important to verify parameters prior to initiating a cycle. Batch reporting by automated systems also facilitate the review of critical parameters. In particular, temperature and relative humidity are important parameters to regularly monitor and trend both in the isolator and the surrounding room during decontamination of isolators.

Steam Sterilisers

PQ for autoclaves have two objectives: (1) to demonstrate the equipment can run a full cycle without alarms, (2) to meet sterility requirements. The standard approach to steam steriliser PQs is to use an overkill validation. This approach requires three consecutive successful half-cycles in order to qualify their proposed full cycle exposure for routine processing of sterilisation loads during commercial manufacturing. A successful cycle typically means all BIs are killed with no growth seen after the incubation period.

For the execution of half-cycle runs, the worst-case loading conditions should be used. Temperature probes, BIs and CIs should be placed throughout the load and include any cold spots or critical locations as documented from cycle development studies. After the test cycle comes to an end, BIs are removed and incubated.

Reference Standards

- ISO 11134 Sterilisation of health care products — Requirements for validation and routine control - Industrial moist heat sterilisation, 1994

- AAMI released the document intended to supersede 11134, with ANSI/AAMI/ISO 17665-1:2006 Sterilisation of health care products — Moist heat — Part 1: Requirements for the development, validation and routine control of a sterilisation process for medical devices (2006)

ISO 17665-1:2006 refers to both product definition and process definition requirements such as product specifications, product families, packaging, re-sterilisation issues, package moisture, stability and potency of container products, re-usable container systems, sterility assurance level/SAL, BIs and CIs, and bioburden determination.

Summary of Requirements

As described above in the suggested IQ/OQ checks, equipment must be qualified (commissioned, IQ/OQ) prior to the validation and performance qualification. Process and cycle development must also be completed before PQ studies.

Thermal Validation Systems

Performance qualification for both depyrogenation tunnels and isolators requires the use of thermal validation systems. These calibrated systems capture temperature data using heat penetrating thermocouples. In advance of conducting a temperature mapping study, the equipment needs to be set up correctly. Most modern systems allow you to define a study in which the temperature range, midpoint and other calibration information is selected. Calibration verification is also completed post-study execution to ensure the accuracy of the data.

Biological Indicators (BIs) for Moist Heat

The biological indicators (BIs) used for the steam sterilisation performance qualifications are typically Geobacillus stearothermophilus. The indicator starting population can be used to calculate the lethality needed to achieve a Sterility Assurance Level SAL of at least 1 x 10-6 is (≥1 x106) and D-value (of 1.5-3.0 at 121°C).

When using an overkill approach, a Spore Log Reduction (aka SLR)/FBIO of ≥15 minutes is required for thermometric measurements, using lethality calculations (F0) during the exposure period of the cycle.

$F0 = D121°C \times (\log No - \log Nf) =$

$F = 1.0 \text{ minute} \times (\log 106 - \log 10\text{-}9)$

$= 15 \text{ minutes}$

No, initial population

Nf, desired final population

Re-Qualification

Depyrogenation: Performance validation for depyrogenation tunnels is normally divided into two separate studies of validation protocols: (1) depyrogenation performance qualification and (2) cool zone sterilisation performance qualification. Annual re-qualification by a completed single run is acceptable to ensure that the equipment is still effective in removing neutralising pyrogens and providing adequate sterilisation.

Isolators: Re-qualification activities are typically limited to conducting one PQ decontamination run on an annual basis. Requalification should also ensure the load pattern and other critical parameters are in control. If amendments are needed such as a change to a loading pattern or quantity of items, then this may require further validation. The impact of the change should be assessed and documented. The purpose of the re-qualification is to confirm the performance of the decontamination cycle is still effective since the initial validation runs.

Culture Media

Culture media is used to fill vials or ampoules in order to evaluate the viable contamination risks from the environment including the personnel. However, culture media is a sterile media, and when processed through a filling and closing machine, will support the growth of microbes if there are issues with the line sterility.

Media fills are studies completed on fill lines in order to demonstrate that sterility can be achieved and maintained during the course of normal manufacturing. Media fills typically are done in advance of process validations.

The Biotechnology HANDBOOK for Engineers

CHAPTER 11

DATA INTEGRITY

Introduction

Data generated by or used in GxP impacting activities must be handled and protected in accordance with international and national regulatory requirements. The application of data integrity applies to many industries and products that touch the lives of patients and end users across the globe. Some examples of products that must meet data integrity regulations include (1) active pharmaceutical ingredients, (2) medical devices, (3) medicinal products, (4) vaccines and (5) cosmetics.

The below agencies and regulatory authorities provide specific requirements on data integrity:

- EU GMP – EudraLex – Rules Governing Medicinal Products in the European Union Volume 4 – Guidelines to Good Manufacturing Practice for Medicinal Products for Human Use – Products for Human and Veterinary Use, Annex 11: Computerised Systems – (1, 7.2, 17)

- FDA – 21 CFR Part 11 – Food and Drug Administration – Electronic Records; Electronic Signatures – Scope and Application (C)

- FDA- 21 CFR Part 211 – Food and Drug Administration – Code of Federal Regulations - Good Manufacturing Practices - 211.188a, 211.194.2, 211.194.8

- ICH E6 – International Conference on Harmonisation - Guideline for Good Clinical Practice (5.2.1, 8.1, 8.3)

- MHRA – United Kingdom - Medicines and Healthcare Products Regulatory Agency - GMP Data Integrity Definitions and Guidance for Industry (2015)

- PIC/S Guidance PI 011-] – Pharmaceutical Inspection Convention Scheme - Good Practices for Computerised Systems in Regulated "GXP" Environments

Within the life science industry the saying goes "if it's not written down, it didn't happen". This is a powerful message that is a suitable starting point for data integrity. In the current and present day, the mere mention of data integrity quickly conjures an image of Excel sheets, big data, databases and computers in our minds. However, it has a broader impact with its roots in the basics of good science – good documentation.

Data integrity indeed does apply to "soft" or electronic data but also applies to paper-based systems and records. GxP is the umbrella acronym that stands of "good practices" in all our tasks and activities, be it laboratory testing, process engineering and so on. A core element in meeting GxP is abiding by "Good Documentation Practices" (GDP). Having good written records is fundamental to patient and product safety within the pharmaceutical, biopharmaceutical and medical devices industries. So, data integrity begins with the small stuff — real-time data collection, real-time review, honest and accurate recording of data and events.

The integrity of data relies on several factors. It can be influenced by a company's culture or approach to doing business. It can also be affected by the level of experience or knowledge within a company. Many traditional engineering companies outside the regulated life science community simply do not have the need to be so thorough in their handling of data and information.

Within a GxP environment, controls, training and the design and operation of systems and processes influence data integrity on a day-to-day basis. Most of the time, those affected by the controls or systems do not think of them, but they can either support or inhibit data integrity and the reliability of data. Obviously, equipment, systems and processes should play a key role in making data reliable and accurate.

Key Terms

Configuration Identification

Software and hardware packages should be identified by a unique product identifier and a version number. For the software end-user, the parts of an automated system that are subject to configuration management should be clearly identified. The system should therefore be broken down into configuration items. These should be identified at an early phase of development so that a complete list of configuration items is defined and maintained. The application-specific items should have a unique name or version ID. The depth of detail when specifying the elements is decided by the needs of the system, and the organisation developing that system.

Requirements for the User ID and Password

User ID: The user ID of a system should have a minimum length agreed with the customer and should be unique within the system.

Password: A password should always consist of a combination of numeric and alphanumeric characters. When setting up passwords, the number of characters and a period after which a password expires should be stipulated. The structure of the password is normally selected to suit the specific customer. The configuration is described in the security settings section of password policy.

Criteria for the structure of a password are as follows:

> ➢ Minimum length of the password
> ➢ Use of numeric and alphanumeric characters

> Case sensitivity

Audit Trail

The audit trail is a control mechanism of a system that allows all data entered or modified to be traced back to the original data. A reliable and secure audit trail is particularly important in conjunction with the creation, change or deletion of GMP-relevant electronic records. In this case, the audit trail must archive and document all the changes or actions made along with the date and time. Typical contents of an audit trail must be recorded and describe the procedures "who changed what and when" (old value/new value).

Data: any data (numerical or otherwise) which is collected or processed as part of GxP activities in order to generate GxP documents and records using a paper-based or electronic process.

Data Handling: Any GxP task that involves creation, entry, review, approval, analysis, reporting, storage, archival, retrieval, or disposal of GxP data.

Data Integrity: The degree to which a collection of GxP data is managed through effective organisational, operational, and technical mechanisms to ensure GxP data reliability.

Data Life Cycle: Starts from the time of data creation to the point of use and during its retention, archival, retrieval and eventual disposal

GxP Impacting: Any action that can impact the quality or safety of a product or critical process.

Application: Software installed on a defined platform/hardware providing specific functionality.

Bespoke/Customised Computerised System: A computerised system individually designed to suit a specific business process.

Commercial Off-the-Shelf Software: Software commercially available, whose fitness for use is demonstrated by a broad spectrum of users.

IT Infrastructure: The hardware and software such as networking software and operation systems, which makes it possible for the application to function.

Life Cycle: All phases in the life of the system from initial requirements until retirement including design, specification, programming, testing, installation, operation and maintenance.

Process Owner: The person responsible for the business process.

System Owner: The person responsible for the availability, and maintenance of a computerised system and for the security of the data residing on that system.

Third Party: Parties not directly managed by the holder of the manufacturing and/or import authorisation.

The Life Cycle of Data

Regulations that speak to GxP and data integrity can apply to many different streams within the life science sector as previously mentioned. From medical devices to pharmaceuticals, all act in different manners, with long and short term applications. Take the example of a total knee replacement. Many designs now ensure their effectiveness in excess of ten years, even up to twenty years depending on individual circumstances. This requires many key records within manufacturing to be kept for several decades. Thus, data retention requirements specify the retention periods of such documents. The integrity of GxP data must be protected during the entire data life cycle, from creation of the data and records to the eventual destruction of data after the retention period is fulfilled. Data integrity equally applies to:

- Equipment
- Computerised systems
- Test records
- Inspection records
- Material certificates

Data integrity ensures that patient safety, product quality, and product supplies are generated by the product life cycle processes.

<u>Process Design</u>

Failure to maintain data integrity can occur throughout the life cycle of data; however, a thoughtful design of systems can prevent breaches in data and restrict the severity of any attempts to alter data. Therefore, design should aim to include controls and preventative measures. At a high level, this can be achieved by:

- Limiting access to GxP events and data
- Standard Operating Procedures (SOPs)
- Training
- System owners

<u>Data Reliability</u>

Data reliability is the foundation to achieving cGxP data integrity. The FDA's ALOCA model can be used to enforce data reliability.

Accuracy: the GxP data is recorded, calculated, analysed, and reported as found and correctly.

Attributable: any actions or calculations performed on GxP data can be attributed to or traceable to the person that performed the actions and the date and time at which they were performed.

Legible: the GxP data is recorded in a clear and human-readable form.

Contemporaneity: the GxP data is recorded at the same time as the observation/measurement is made or as soon as possible after the event.

Original: the initial data recorded is available and not altered.

An additional point to make it that of trustworthiness. It is assumed that engineers and scientists etc. working across the life science industries are ethical and do not falsify data or information. Typically companies can implement a code of practice or ethical behaviour programme to desist people from intentional unethical behaviour or the falsification of records.

Data Creation: The point at which the values or data is created. The data and information is original (raw).

Data Authentication: Within a GxP environment, authentication refers to the approval of data (electronic signatures). E-signatures are key controls within software that prompt the user to enter a unique username and password to acknowledge a recording or action. The e-signature should create a permanent link with the electronic record that cannot be removed and can be viewed through an audit trial.

Data Protection: Once the data is created, the handling of the data must ensure data integrity. For electronic data, this includes access control to computer systems. Other practical restrictions can also be made such as limiting room and site access to authorised personnel.

Data Retention: This refers to the controlled storage, backup and arching of data. Retention of records may be required for several decades depending on the type of data and the regulatory requirements relating to the particular product or industry.

Technical Controls

The benefits of modern software and computerised systems allow robust and complex data handling and calculations to be completed. With this modern capability that is becoming more powerful comes more responsibility with regard to the use of data.

The computerised systems used to generate, gather or interpret GxP data must fulfil several criteria. First and foremost, they must be fit for the intended use. The software and hardware must be validated and proven to be consistent and reliable. Some general considerations for the use of computerised systems include:

- Systems designed to foster integrity of GxP data
- User requirements specification detailing the intended use and required functionality
- An approved vendor with certification to ISO 9001 or other quality management standards
- Software should meet the requirements of regulations such as FDA 21 CFR Part 11
- Written procedures on how automated processes function

It should not be an easy process for personnel to alter or corrupt data when using computerised systems. GxP-impacting computer systems should have controls that prevent unauthorised access along with audit trail history.

Audit trail design and configuration capture key critical processes, events, settings and information. This enables any investigations of quality events impacting data integrity to be reviewed and analysed.

Computer System Design and Development

For computer systems, software requirements are typically stated in functional terms and are defined, refined and updated during the development phase. Success in accurately and completely documenting software requirements is a crucial factor in successful validation of the resulting software. A specification* is defined as "a document that states requirements." It may refer to or include engineering drawings or other relevant documents *21 CFR 820.3(y).

There are different kinds of written specifications:

- User requirements specifications
- System requirements specification
- Software requirements specification
- Software design specification
- Software test specification
- Functional design specification

All of these documents establish "specified requirements" and are design outputs for which various forms of verification or validation are required. The URS must also define non-software requirements and hardware. Non-functional requirements such as maintainability and usability can also be included. There should be a clear distinction between mandatory regulatory requirements and optional features. Proper definition at this stage ensures the system meets data integrity requirements and prevents costly updates down the line.

Practical Elements to Data Integrity

Facilities and systems must be configured in a way that encourages compliance with principles of data integrity. Examples include:

- Availability of clocks for recording times.
- Access points to allow swift reference to GxP records at locations where tasks are completed.
- Control of raw data.
- Control of approved documents.

Organisational Controls

Regulated companies such as medical device, pharmaceutical and biotechnology companies are required to operate under a quality management system. For medical devices, ISO 13485 serves as a quality management system. Likewise, the FDA Code of Federal Regulations 21 CFR Part 211 functions as a QMS for finished pharmaceuticals.

Organisational controls for Data Integrity can address:

- Assessment of GxP computerised systems
- Management of GxP computerised systems
- Electronic Records Implementation and handling
- Use of Electronic signatures
- Quality Risk Management

Operational Factors

Operational factors refer to process or manufacturing errors, deviations or non-compliance to established procedures that may impact data integrity.

GxP data handling activities should be designed to limit human intervention. As with human intervention there can be errors or omissions. Furthermore, it may call into question the reliability of the data.

Mistake-proofing methodologies should be developed to avoid human error related breaches in data integrity. As with any system or technology, training is a fundamental step. Building upon training, exposure to GxP data systems and on-the-job training all play a part in delivering a system that is robust and meets regulatory requirements. It is important to remind ourselves that while regulations are the driving force to comply with data integrity, the ultimate goal is always the protection and safety of the patient or end user of the product, medicine or treatment.

Software Validation

Where there is the potential to affect product conformance to requirements or where software or IT systems provide support to aspects of quality management, validation is required. Most companies categorise software validations to account for the different applications of software and IT systems. For example, enterprise systems, such as the drawing package SolidWorks would be validated in a different manner to manufacturing systems that contain software (a.k.a. embedded software).

"Embedded" software is where the software is integrated into the manufacturing equipment. Embedded software is typically validated during the equipment qualification stage, process validation stage or test method validation. Enterprise software falls outside of equipment or process validation but does require validation if it impacts product quality or is used to make quality decisions. Standalone systems such as ERP (Enterprise Resource Planning) systems also require validation.

Software Validation and GAMP

Good Automated Manufacturing Practice (GAMP) is a set of guidelines for manufacturers and users of automated systems in regulated industries. GAMP specifically impacts the medical device, pharmaceutical and biopharmaceutical industries. The application of GAMP and validation of automated systems in manufacturing helps ensure that regulated medical devices and medicinal products have the required quality and are manufactured according to good practices, meet regulatory and legal requirements and ensure patient safety. GAMP ensures quality is in-built into each stage of the manufacturing process. Therefore, GAMP has a place in all aspects of automation and production, including the handling of raw materials, control of facilities and equipment etc.

Key Terms

Automated System: Term used to cover a broad range of systems, including automated manufacturing equipment, control systems, automated laboratory systems, manufacturing execution systems and computers running laboratory or manufacturing database systems. The automated system consists of the hardware, software and network components, together with the controlled functions and associated documentation. Automated systems are sometimes referred to as computerised systems; in this guide the two terms are synonymous.

Commercial Off-the-Shelf (COTS): Configurable programs and stock programs that can be adapted to specific user applications by "filling in the blanks" without (COTS) altering the basic program.

Computer System Validation: A process that confirms by examination and provision of objective evidence that the computer system conforms to user needs and intended uses. System validation is a process for achieving and maintaining compliance with GxP regulations and fitness for intended use by adoption of life cycle activities, deliverables, and controls.

GAMP 5: A set of guidelines that offers a risk-based approach to ensuring the compliance of GxP-impacting computerised systems.

V- Model: A development process which sets out a roadmap of stages and deliverables during a project.
21 CFR Part 820: FDA requirements pertaining to medical devices.

User Requirement Specification, URS: The URS is a critical document that defines the requirements of the computerised system and agreement to the requirements.

Software Requirement Specification, SRS: An SRS can be written to interpret the requirements of a URS and how they relate to the requirement or how the requirement is met in practical terms regarding software.

Functional Design Specification, FDS: A functional design specification is a document that specifies how particular requirements are met — this can be a combination of how the equipment/process operates mechanically/automatically etc. An FDS is typically written in response to a URS.

Computer System Validation Life Cycle

The computer system validation life cycle refers to all activities from initial concept to retirement of a computer system. The life cycle of the system includes the defining of, and performance of activities in a systematic way from conception, requirements, development or configuration, testing, release and operational use. The four GAMP life cycle phases include:

- Concept
- Planning and project stage
- Operation
- Retirement

The concept stage is concerned with understanding the need or the problem to be addressed. We will see that the user requirement specification (along with other specifications) and the initial risk assessment help to drive a project forward in a systematic manner. The most common life cycle approach for computerised and automated systems is the V-Model. The GAMP-based V-model lays out a roadmap which facilitates the validation of equipment and automated systems.

The planning and project stage involves the planning of the validation effort required to implement the system into the business area(s) based on identification and approval of system concept. This phase includes assessments of the regulatory and system risks, supplier assessment, development of validation strategies, identification of deliverables that will be generated, definition of the business process the system will support as well as the user requirements which the system will fulfil.

Design, development and configuration of the hardware and software is also required to meet the system requirements as per specifications. In the case of custom software components, this effort could also include detailed software design and developmental testing to ensure readiness for verification testing.
The verification stage confirms that specifications have been met and releases the system for use. This phase will involve multiple stages of reviews and testing depending on the system type, the development method applied and its use. Once verification activities have begun, any changes to the system must be captured through change control.

On successful completion of the verification activities, the system is then released for effective use. The test strategy and other verification activities will vary widely between simple equipment and more complex customised/configurable systems. The verification and validation approach is typically agreed and detailed at the validation planning stage. The VP can be updated accordingly as the project develops with more detail being added. Alternatively, a test strategy document or matrix could be written to provide more specific test plans.

Verification deliverables vary based on the complexity and level of customisation of the system in question. Corporate or company specific procedures also shape the required activities to be completed and reported. Some generic deliverables are listed below.

- Approval, execution and review of test protocols
- Writing and approving SOPs for operation and maintenance of the system
- Traceability matrix
- Completion of any risk mitigations (e.g. updates to FMEA etc.)
- Validation summary report(s)

Validation reporting requirements vary depending upon the scope of the system and should also be driven by a procedure and template. The validation plan can also outline the deliverables and what needs to be addressed in the report. A Validation Summary Report (VSR) should be written to summarise the results of executing the VP, the documents created for the validation activities and the testing performed. Finally, the VSR indicates the acceptance of the system/equipment by the user and the project team and states that the equipment is released for commercial operation/production.

The operation phase supports the need to maintain compliance and fitness for intended use after the system is released for normal use. It is important to ensure the system remains within a continued validated state. All proposed or necessary changes to the system must be assessed and controlled as part of a change control process. Once the system has been accepted and released for use, the operation phase begins. This phase consists of maintaining the system's compliant state and fitness for intended use through the control of the procedures supporting the system's operational use.

During the operation phase, the below activities are typically completed:

- Ongoing training
- Preventative maintenance
- Service management and performance monitoring.
- Change control
- Periodic review
- Maintaining system security
- Records management
- Calibration

The retirement phase involves the planning and proper management of activities relating to the removal of systems from service (shutdown). The retirement should take into account the storage of any data and any data migration that needs to occur prior to retirement. The retirement plan, if needed, will outline the retirement strategy from the roles and activities that will be conducted to the removal of the system for use. A retirement summary report is produced that documents the results of the activities defined in the retirement plan including:

> ➢ Retirement plan and timelines.
> ➢ Summaries of any data migration activities.
> ➢ Identification of the storage location of documentation relating to the system.
> ➢ Obsoleting of SOPs.

It must be stressed that GAMP is a set of principles, a set of guidelines that aim to achieve compliant computerised systems that are fit for intended use. GAMP guidelines differ to 21 CFR QSR regulations as they are not legal or statutory requirements. However, they represent industry best practice and complement the validation efforts that are legal requirements and statutory requirements.

Regulatory Review

Software validation is a requirement of the quality system regulation, 21 Code of Federal Regulations (CFR) Part 820. Validation requirements apply to:

(1) software used as components in medical devices,
(2) software that is itself a medical device, and
(3) software used in production of the device or in implementation of the device manufacturer's quality system.

Note: EU GMP Annex 11, provides information on the inspection of 'Computerised Systems'.

In addition, computer systems used to create, modify, and maintain electronic records and to manage electronic signatures are also subject to the validation requirements. Such computer systems must be validated to ensure accuracy, reliability, consistent intended performance, and the ability to discern invalid or altered records. The regulated user should be able to demonstrate through the validation evidence that they have a high level of confidence in the integrity of both the processes executed within the controlling computer system and in those processes controlled by the computer system within the prescribed operating environment.

System Categorisation

GAMP 5 makes provision for four categories of software in order to distinguish the level of tcustomisation/configurability that exists across software serving different functions:

GAMP Software Category 1, Operating Systems
GAMP Software Category 2, Non-configured software
GAMP Software Category 4, Configurable software packages
GAMP Software Category 5, Custom Software

GAMP Software Category 1, Operating Systems

Category 1, operating systems, covers established commercially available operating systems. These systems are not subject to validation themselves. The name and version of the operating system must, however, be documented and verified during Installation Qualification (IQ). Application software hosted on operating systems needs to be validated.

GAMP Software Category 3, Non-Configured Software

Category 3 covers commercially available, standard software packages and "off the-shelf" solutions for certain processes. The configuration of the software packages should be limited to adaptation to the runtime environment (for example network and printer connections) and the configuration of the process parameters. The name and version of the standard software package should be documented and verified in an installation qualification (IQ). Special user requirements, such as security, alarms, messages, or algorithms must be documented and verified in an operational qualification (OQ).

GAMP Software Category 4, Configurable Software Packages

GAMP Software Category 4, Configurable Software Packages Category 4 covers configurable software packages that allow special business and manufacturing processes. This involves configuring predefined software modules. These software packages should only be considered as belonging to Category 4 if they are well-known and mature. Normally, a supplier audit is necessary. If this is not available, the software packages should be handled as Category 5. The name, version, and configuration should be documented and verified in an installation qualification (IQ). The functions of the software packages should be verified in terms of the user requirements in an operational qualification (OQ). The validation plan should take into account the life cycle model and an assessment of suppliers and software packages.

GAMP Software Category 5, Custom Software

GAMP Software Category 5, Custom Software Custom/Bespoke Software (GAMP Software Cat 5) is software that contains custom code designed or modified specifically for a particular customer. As the code is custom, it presents a greater risk. This risk must be mitigated with the right approach to the validation.

GAMP Considerations

Correctly assigning a GAMP software category to equipment, systems or processes is an important activity that should be completed early on in the planning stage of a project. There must of some degree of familiarity with the equipment or system. The manufacturer or vendor can be a source of information that may help the designation. In many cases, companies create tools or processes that help determine what GAMP software category applies. These have different names such as questionnaires, screening tools, planning tools etc.

Risk Assessments

A risk assessment process should be applied to cGxP computerised systems in order to identify and mitigate potential risks to (1) patient safety, (2) product quality and (3) data integrity. Results identified through a risk assessment help to determine the validation strategy, the effort and time required, and allow better targeting of the validation activities to the highest risks.

The risk assessment should be revised during the software development lifecycle (SDLC) if the functionality, requirements or intended use of the system changes. The risk assessment activity should also be evaluated during system build-up as well as when implementing changes. Risk assessment tools for cGxP computerised systems are typically completed during the planning stage, specification stage and post-qualification if a change or update is required.

Planning Stage

Initial Impact/Risk Assessment – takes place during the planning phase to identify the level of impact and GxP relevance of the system/equipment. (Tools used: High Level Risk Assessment).

Specification Stage

Functional or Quality Risk Assessment – takes place during the specification phase and identifies potential risks and possible mitigations to be to be introduced to the process. (Tools used: Quality Risk Matrix, (p)FMEA).

Changes to the System

Impact Assessment of Changes – takes place as part of the change control process in the system operational phase.

Quality Risk Matrix

A QRM is a risk assessment that identifies and manages the risk to patient safety, product quality and data integrity that relate to system processes. Risk scenarios or potential causes should be developed for each identified function or process step and then assessed for the impact on patient safety, product quality or data integrity. Risk mitigations and controls should then be introduced to address both medium and high levels of risk. The QRM requires three "assessments" in order to produce an estimation or overall risk (low, medium, high),

- ➢ Assess likelihood
- ➢ Assess detectability
- ➢ Assess severity

Traceability Matrix

A traceability matrix should be prepared as required in accordance with company and internal policy. It is also recommended by GAMP guidelines, ASTM E2500 and ISPE risk-based approaches to validation. The matrix links the user requirements and specifications to testing and validation activities. A traceability matrix illustrates that all user requirements are traceable to the verification/validation activity or vendor documents as relevant (FDS if applicable, design specifications etc.) Generally, individual organisations will have an approved template to work from. However, the URS structure can form the basis of the template, with additional columns added to document the test/verification method or reference documents (such as FDS and vendor specifications and design documents)

21 CFR Part 11

This section specifically covers the regulatory requirements of part 11 of Title 21 of the Code of Federal Regulations; Electronic Records; Electronic Signatures (21 CFR Part 11). Part 11 of the FDA CFR is relevant to "records in electronic form that are created, modified, maintained, archived, retrieved, or transmitted under any records requirements set forth in agency regulations."

As of 2007, several sections of the regulation have been identified as excessive and the FDA announced in guidance that it will exercise enforcement discretion on some parts of 21 CFR part 11. This has been welcomed by some manufacturers but it has also caused a degree of confusion. The requirements relating to access controls are the most fundamental requirements and are routinely enforced. The "predicate rules" that required organisations to keep records in the first place are still in effect. If electronic records are illegible, inaccessible, or corrupted, manufacturers are still subject to those requirements.

If a regulated firm keeps "hard copies" of all required records, those paper documents can be considered the authoritative document for regulatory purposes. This then means that the computer system is not in scope for electronic records requirements, although subject to predicate rules which still require validation. If the "hard copy" is to be identified as the authoritative document, the "hard copy" must be a complete and accurate copy of the electronic source. The manufacturer must use the hard copy (rather than electronic versions stored in the system) of the records for regulated activities.

Definition of Records

The FDA has deemed the following records or signatures in electronic format subject to 21 CFR part 11:

Records that are required to be maintained under predicate rule requirements and that are maintained in electronic format in place of paper format. On the other hand, records (and any associated signatures) that are not required to be retained under predicate rules, but that are nonetheless maintained in electronic format, are not part 11 records.

Records that are required to be maintained under predicate rules, that are maintained in electronic format in addition to paper format, and that are relied on to perform regulated activities.

Records submitted to FDA, under predicate rules (even if such records are not specifically identified in agency regulations) in electronic format (assuming the records have been identified in docket number 92S-0251 as the types of submissions the agency accepts in electronic format). However, a record that is not itself submitted, but is used containing nonbinding recommendations in generating a submission, is not a part 11 record unless it is otherwise required to be 205 maintained under a predicate rule and it is maintained in electronic format.

Electronic signatures that are intended to be the equivalent of handwritten signatures, initials, and other general signings required by predicate rules. Part 11 signatures include electronic signatures that are used, for example, to document the fact that certain events or actions occurred in accordance with the predicate rule (e.g. approved, reviewed, and verified).

The above definitions are taken from the FDA guidance document entitled "FDA Guidance for Industry: 21 CFR Part 11 - Electronic Records and Electronic Signatures: Scope and Application, August 2003." This document also provides recommendations on documenting key decisions that may be taken in relation to 21 CFR Part 11 applicability and compliance.

Requirements and Specifications

The need for compliance to 21 CFR depends on the type of technology and level of automation and computerisation involved in the manufacturing process or other actives that are GxP-impacting. Does the system store electronic records? Does the system require a login? Is there an audit trial? If a complex system is to be procured, the requirements need to be communicated to the manufacturer as part of a user requirement specification and/or software requirement specification.

General Guidance on Requirement Specifications

While the quality system regulation states that design input requirements must be documented, and that specified requirements must be verified, the regulation does not further clarify the distinction between the terms "requirement" and "specification." A requirement can be any need or expectation for a system or for its software. Requirements reflect the stated or implied needs of the customer, and may be market-based, contractual, or statutory, as well as an organisation's internal requirements.

There can be many different kinds of requirements (e.g., design, functional, implementation, interface, performance, or physical requirements). Software requirements are typically derived from the system requirements for those aspects of system functionality that have been allocated to software. Software requirements are typically stated in functional terms and are defined, refined, and updated as a development project progresses. Success in accurately and completely documenting software requirements is a crucial factor in successful validation of the resulting software. *Page 6 Guidance for Industry and FDA Staff General Principles of Software Validation A Specification* is defined as "a document that states requirements." (21 CFR 820.3(y)). It may refer to or include drawings, patterns, or other relevant documents and usually indicates the means and the criteria whereby conformity with the requirement can be checked.

There are many different kinds of written specifications, e.g., system requirements specification, software requirements specification, software design specification, software test specification, software integration specification, etc. All of these documents establish "specified requirements" and are design outputs for which various forms of verification are necessary.

Validation of Computerised Systems

The requirement for computerised systems to be compliant to 21 CFR part 11 needs to be identified early on in the project to ensure that the vendor or supplier of the systems or equipment can develop and build a system that meets the requirements of 21 CFR part 11. Computer system validation can be divided into three distinct phases: (1) planning, (2) design and development, (3) verification and (4) retirement.

Planning: This phase involves the planning of the validation effort required to implement the system and identification of key milestones and requirements. It requires supplier assessments, assessments of the regulatory and system risks, supplier development of a validation approach and the identification of deliverables that will be generated to support the implementation and operation of the system.

Design and Development: This phase consists of the design, development and configuration of the hardware and software required to meet the system requirements. In the case of custom software, design and developmental testing is important to ensure proper functionality prior to verification testing.

Verification: This phase confirms that requirements and specifications have been met. Testing is required to ensure the system operates as intended. Upon successful testing and verification, the system can be released for use. Once verification activities have begun, any changes to the system must be managed through change control. In case of successful completion of the verification activities (i.e. any deviation has been evaluated and addressed), the system is released for effective use. The operation phase supports the need to maintain compliance and fitness for intended use after the system is accepted and released for use.

Retirement: This phase consists of the planning, executing and summarising of the events required for system shutdown. It includes the appropriate handling of the supporting documents and the data contained within the system. While described here as a separate phase, a system's retirement can be handled as part of a new system implementation or as a separate project.

Best practice when it comes to computer system validation is to adopt a life cycle approach which requires the completion of activities in a systematic way from system conception to retirement. Life cycle activities could be scaled according to system impact on product quality, patient safety and data integrity, system complexity and novelty, supplier assessment and business risk.

Definitions

Computer System: A computer/automated system consisting of the hardware, software, and network components, together with the controlled functions (personnel, procedures, and equipment) and associated documentation.

Computer System Validation: A process that confirms by examination and provision of objective evidence that the computer system conforms to user needs and intended uses. Computer system validation is a process for achieving and maintaining compliance with GxP regulations and fitness for intended use by adoption of life cycle activities, deliverables, and controls.

GxP-Regulated Computer Systems: Computer systems determined to have a potential impact on product quality, patient safety and data integrity; these systems are required to comply with the relevant GxP regulations.

Data Integrity: The degree to which data is reliable and without error. Data must be accurate, attributable, contemporaneous, original, legible and available. A breach of data integrity occurs when any person manipulates or distorts data and submits the results of that data as valid.

Predicate Rules: A predicate rule is any FDA regulation that requires companies to maintain certain records and submit information to the agency as part of compliance.

To gain a better understanding of the validation of computerised systems, consult the following publication: "FDA's Guidance for Industry and FDA Staff General Principles of Software Validation." See also industry guidance such as the GAMP 5 guide issued by ISPE for a useful reference.

Electronic Records

When it comes to the regulated industries such as the medical device industry, every process and procedure must be documented. Documentation ensures that everyone is working in the same manner with the same procedures. However, documentation is more than just writing down procedures and processes. It is also concerned with how documents are controlled, how they are updated and how they are stored.

Electronic Document management systems

Electronic document management systems aka EDMS are now the norm and gold standard for most medium to large organisations. Many companies that provide medical device manufacturers with an EDMS that can be customised to match the business processes particular to an organisation. With configurable or customisable software, validation and proper verification is important to ensure the system operates as intended. There are also regulatory requirements that stipulate the expectations and requirements of such systems. For example, the application of electronic signatures and the presence of audit trials. FDA 21 CFR Part 11 details the requirements with regard to electronic records and electronic signatures. For medicinal products in Europe, GMP V4 Annex 11 specifies similar requirements.

Record Retention

With regard to the part 11 requirements for the protection of records to enable their accurate and ready retrieval throughout the records retention period (11.10 (c)), persons must also comply with all applicable predicate rule requirements for record retention and availability such as (211.180(c) general requirements. The decision to follow 21 CFR part 11 should be justified and documented as part of a risk assessment and based on the value of the records over time.

The FDA does not object to archiving of required records in electronic format to non-electronic media such as paper, or to a standard electronic file format (examples of such formats include, but are not limited to, PDF, XML, or SGML). Persons must still comply with all predicate rule requirements, and the records themselves and any copies of the required records should preserve their content and meaning. As long as predicate rule requirements are fully satisfied and the content and meaning of the records are preserved and archived, you can delete the electronic version of the records. In addition, paper and electronic record and signature components can coexist as long as predicate rule requirements are met and the content and meaning of those records are preserved.

Electronic Signatures

Electronic signatures are computer-generated character strings that count as the legal equivalent of a handwritten signature. The regulations for the use of electronic signatures are set out in 21 CFR Part 11 of the FDA. Each electronic signature must be assigned uniquely to one person and must not be used by any other person. It must be possible to confirm to the authorities that an electronic signature represents the legal equivalent of a handwritten signature. Electronic signatures can be biometrically based or the system can be set up without biometric features.

Conventional Electronic Signatures

If electronic signatures are used that are not based on biometrics, they must be created so that persons executing signatures must identify themselves using at least two identifying components. This also applies in all cases in which a chip card replaces one of the two identification components. These identifying components, can, for example consist of a user identifier and a password. The identification components must be assigned uniquely and must only be used by the actual owner of the signature.

When owners of signatures want to use their electronic signatures, they must identify themselves by means of at least two identification components. The exception to this rule is when the owner executes several electronic signatures during one uninterrupted session. In this case, persons executing signatures need to identify themselves with both identification components only when applying the first signature. For the second and subsequent signatures, one unique identification component (password) is then adequate identification.

Audit Trail

Title 21 CFR details predicate rule requirements relating to documentation of, for example, date time, or the sequencing of events, as well as any requirements for ensuring that changes to records do not obscure previous entries. Making the decision on whether to apply audit trails, or other appropriate measures, or on the need to comply with predicate rule requirements should involve a justified and documented risk assessment. Any risk assessment should determine the potential effect on product quality and safety and the integrity of the record.

Change Management

Validation programmes are subject to change control. Each company or organisation should have a procedure detailing the change management process.
Any system, facility, document or process that has the potential to impact product quality and the validated state is generally subject to following a change control process. Another term used in industry is enterprise change control or engineering change control. Essentially, these terms are the same. The intent is to control

and manage change consistently.

A change control can take the form of a document which drives the agenda and the specific requirement. Change control is also created with enterprise software such as Kintana, Documentum and SAP. While each company will have varying processes, some basics are common. These include the three stages of change control; pre-implementation, implementation and post implementation (if required).

Validation Deliverables

The deliverables of validation activities should be in accordance with a project validation plan of validation master plan. For small projects or changes to computerised systems, a change control may serve as the validation plan. However, some typical deliverables include the following:

- GxP assessment (note, some systems may be non GxP applicable)
- User requirements specification
- Third party audit
- Validation plan
- Design specification such as functional, software, hardware and technical specifications
- GxP risk assessment
- Validation protocols
- Traceability matrix
- Validation report

GLOSSARY

A

Accelerated Ageing

When the deterioration of a device or product component from natural ageing is accelerated and simulated in the laboratory.

Accuracy

Accuracy or trueness. An expression of the closeness of agreement between the value that is accepted, either as a conventional true value or an accepted reference value and the value obtained. A system with low bias implies good accuracy and vice versa.

Adverse Event

A situation or condition that occurs when a data point, result, or process etc. is outside the expected or predetermined limits or ranges.

Air Exchange Rate per Hour (ACPH)

The rate of air exchange expressed as the number of air changes per hour and calculated by dividing the volume of air delivered in the unit of time by the volume of space.

Active Pharmaceutical Ingredient

Any substance or mixture of substances intended to be used in manufacturing a drug (medicinal) product and that, when used in the production of a drug, becomes an active ingredient of the drug product. Such substances are intended to furnish pharmacological activity or other direct effect in the diagnosis, cure, mitigation, treatment, or prevention of disease, or to affect the structure and function of the body. (ICH Q7A, Annex 18, Part II.)

ANSI

American National Standards Institute

Antimicrobial Resistance

Antimicrobial resistance corresponds to the emergence and spread of microbes that are resistant to cheap and effective first-choice, or "first-line" antimicrobial drugs.

Application

A term most often used in relation to software validation and computerised systems. It is any software installed on a defined platform providing specific functionality.

Approve

"Approve" means green-lighting the device after reviewing a premarket approval (PMA) application that has been submitted to FDA.

AVL (Approved Vendor List)

A list of all the vendors or suppliers approved by a company as sources from which to purchase materials.

Artwork

Electronic files or printouts containing the representation of a packaging item, graphical elements, and regulatory text. Approved artworks are used by suppliers for printing.

Aseptic (Conditions)

Conditions in the working environment under which the potential for microbial and/or viral contamination is minimised.

ASTM

Acronym for The American Society for Testing and Materials.

ATEX

An acronym derived from the French-titled 'Atmosphères Explosibles' 94/9/EC directive outlining what equipment and work environment is allowed in an environment with an explosive atmosphere. This European directive amends and adds safety requirements for hazardous areas in the relevant national legislation in the member states of the European Union, bringing in a common standard. Where equipment is to be used in potentially explosive atmospheres containing gas or combustible dust, it must comply with the ATEX directive.

Audit Trail

The audit trail is a control mechanism of a system that allows all data entered or modified to be traced back to the original data. A reliable and secure audit trail is particularly important in conjunction with the creation, change or deletion of GMP-relevant electronic records.

Acceptable Quality Level (AQL)

The AQL of a sampling plan is the Process Performance Level routinely accepted by the sampling plan.

B

Basis of Design

A design document that demonstrates a thorough understanding of the project and its intended output. Typically contains preliminary drawings and system descriptions etc. Together with the URS and the detailed design, it provides overall evidence that the design addresses the requirements of the equipment, system or facility.

Biocompatibility

A measure of how a biomaterial interacts in the body with the surrounding cells, tissues and other factors.

Bioburden

The level and type of micro-organisms that can be present in raw materials, API starting materials, intermediates or APIs. Bioburden should not be considered contamination unless the levels have been exceeded or defined objectionable organisms have been detected.

Biological Indicators

A test system containing viable microorganisms providing a defined resistance to a specified sterilisation process, e.g. vaporised hydrogen peroxide.

Biomaterial

Any matter, surface, or construct that interacts with biological systems. Biomaterials can be derived from nature or synthetic (manufactured). The active substance of a biosimilar medicine is comparable to a biological reference medicine. Biosimilar and biological reference medicines are used at the same dose to treat the same disease. The name, appearance and packaging of a biosimilar medicine differs to that of a biological reference medicine.

Bracketing

A bracketing (aka family or matrix) approach can be used where similar products are produced using the same equipment and processes. A particular product size or product configuration may be selected to represent the worst-case product. Therefore, by qualifying the worst case, all of the other products within the family are considered validated.

Body Orifice

Any natural opening in the body, as well as the external surface of the eyeball, or any permanent artificial opening, such as a stoma or permanent tracheotomy.

Borderline Classifications

In certain circumstances, it may not be clear if a product falls under the medical device legislation or whether to classify a device as a medicine, cosmetic, biocide and so on. The decision will largely depend on the particular intended use of the product, as assigned by the manufacturer, and on the demonstrated mode of action. The manufacturer's claims must be substantiated by relevant data.

Bulk Product

Any pharmaceutical form (liquid, powder, suspension) that is to be filled into either another container or its final container at the next process step or is already filled into its final container to be labelled and packaged at the next process step.

BOM

Bill of Materials.

BSI

British Standards Institute.

C

CAD, Computer Aided Drawing

A system used to create physical designs, usually three-dimensional. Some examples of CAD software are SolidWorks, Pro/ENGINEER and AutoCAD.

Calibration

A requirement that demonstrates a particular instrument or device produces results within specified limits by comparison with those produced by a reference or traceable standard over an appropriate range of measurements.

Campaign (Process)

A production strategy where consecutive batches of an API, a finished product, or intermediates are processed before the production line/system is cleaned.

Capability (Process Capability)

Process capability is a measure of how capable the process is of producing product meeting specified requirements. It is a measure of the actual variation in that product characteristic compared to the product specifications. Indices are used to represent the process capability such as Pp, Cp and Ppk, Cpk, depending on how the data is collected, e.g. multiple batches over time.

CAPA

A **c**orrective **a**nd **p**reventive **a**ction. A systematic approach that includes actions needed to correct, prevent recurrence and eliminate the cause of potential nonconforming product and other quality problems (preventive action) (21CFR 820.100).

Change Control

A formal system by which qualified representatives of appropriate disciplines review proposed or actual changes that may impact the validated status.

Change Notification (Agreement)

A signed declaration that states that the supplier agrees to notify the customer of changes in its product or process in order to allow the customer determine whether the changes can affect the quality of finished goods or quality system.

Change Management

An overarching approach to change control that is used during the preliminary planning and design stage of a project.

Cleaning

The process of removing potential contaminants from process equipment and maintaining the condition of equipment so that the equipment can be safely used for subsequent product manufacture.

Cleaning Validation

Documented evidence that provides a high degree of assurance that a specific cleaning process will consistently produce a result meeting predetermined requirements for cleanliness.

Cleaning Verification

Confirmation by examination and provision of objective evidence that specific requirements have been fulfilled.

Cocurrent (Flow)

This is when the fluids are applied in the same direction. Cocurrent flow is less effective as less heat can be transferred, therefore it is less commonly used.

Code of Federal Regulations (CFR)

Regulations issued by U.S. government agencies. The individual titles making up the regulations are numbered the same way as the federal laws on the same topic.

Competent Authority

A competent authority is the legally designated authority mandated to monitor compliance with directives and legal requirements within the industry. The competent authority has the power to grant and revoke licenses.

Compendial Organisations

Organisations certifying material standards that meet compendial requirements and acceptance criteria, e.g. the United States Pharmacopeia.

Commissioning

An engineering activity that includes all aspects of introducing a system, piece of equipment or process is installed and ready for use. Commissioning involves both requirements of installation qualification (IQ) and operational qualification (OQ).

Computer System

A group of hardware components and associated software, designed and assembled to perform a specific function or group of functions. [EU GMP Guide, Part II, ICH Q7.]

Computerised System

A system including the input of data, electronic processing and the output of information to be used either for reporting or automatic control. [EU GMP Guide, Glossary.]

Computer System Validation

A process that confirms by examination and provision of objective evidence that the computer system conforms to user needs and intended uses. System validation is a process for achieving and maintaining compliance with GxP regulations and fitness for intended use by adoption of life cycle activities, deliverables, and controls.

Concurrent Validation

Concurrent validation occurs when activities are executed at the same time as one another or concurrent to a product launch.

Confidence Level

Confidence level is expressed as a percentage and represents the probability that the conclusion of the test is correct. A 95% confidence level means you can be 95% certain that the conclusion is correct.

Conflict of Interest

A conflict of interest is a situation in which a public official's decisions are influenced by the official's personal interests.

Continual Improvement, CI

Ongoing activities to evaluate and positively change products, processes and the quality system to increase effectiveness

Consent Decree

A consent decree is a binding order issued by a judge that stipulates the voluntary agreement by the participants in a case of litigation. Decrees are sometimes issued after one party voluntarily agrees to cease a particular action without admitting to any illegality of the action to date.

Colony Forming Unit

One or more microorganisms that produce a visible, discrete growth on an agar-based microbiological medium.

Controlled Substances

Products that are categorised due to their potential for abuse, medical use and requirement for medical supervision.

Controlled Classified Areas

An environment supplied with HEPA-filtered air where materials, equipment, and personnel are regulated to control viable and non-viable particulates to an acceptably low level. Such areas are classified according to the maximum level of airborne particulate allowed.

CNC (Controlled Not Classified)

While these are not ISO-recognised room classes, they are generally used to describe non-GMP areas with a level of control in effect.

Clear (FDA)

The attainment of FDA 'clearance' for the device after reviewing a premarket notification, otherwise known as a 510(k) (named after a section in the Food, Drug, and Cosmetic Act) that has been filed with FDA.

Clean Room

An area (or room or zone) with defined environmental control of particulate and microbial contamination, constructed and used in such a way as to reduce the introduction, generation and retention of contaminants within the area.

Containment

A process or device to contain product, dust or contaminants in one zone, preventing it from escaping to another zone.

Contamination

The undesired introduction of impurities of a chemical or microbial nature, or of foreign matter, into or onto a starting material or intermediate, during production, sampling, packaging or repackaging, storage or transport.

Continued Process Verification

Once the initial validation is completed it is important that the system or process remains within the validated state. This is done by monitoring the performance and output of the system or equipment. Furthermore, any changes to this system or equipment must be assessed and documented in order to ensure the product is safe and meets acceptance criteria.

Critical Aspects

Critical aspects of manufacturing systems include the functions, features, abilities, and performance or characteristics required for the manufacturing process and systems to ensure consistent product quality and patient safety. They should be identified and documented based on scientific product and process understanding.

Critical Quality Attribute, CQA (Critical-to-Quality)

A property or characteristic with specific nominal value and appropriate limit and range providing a particular quality attribute. A CQA is typically classed as a high-risk requirement, where the safety or efficacy of the product depends on the CQA being within the specified limits.

CCC (Mark)

The China Compulsory Certificate mark, commonly known as a CCC Mark, is a safety mark for many products sold on the Chinese market. As of 2013, medical devices do not require this certification.

CDC

Centre for Disease Control & Prevention (USA).

CDRH

Centre for Devices and Radiological Health (USA).

CE Marking

CE Marking is a mandatory conformance mark on many products (including medical devices) placed on the single market in the European Economic Area. The CE marking certifies that a product has met EU consumer safety, health or environmental requirements. By affixing the CE marking to a product, the manufacturer declares that it meets EU safety, health and environmental requirements.

CEN

Communité Européenne des Normes (European Committee for Standardisation).

Clinical Trial

Clinical trials are conducted to allow safety and efficacy data to be collected for health interventions (e.g. drugs, diagnostics, devices, therapy protocols). These trials can only take place after satisfactory information has been gathered on the quality of the non-clinical safety, and health authority/ethics committee approval is granted in the country where the trial is taking place.

Clinical Trial Sponsor

The clinical trial sponsor is responsible for the safety of subjects in a clinical trial and informs local site investigators of the true historical safety record of the drug, device or other medical treatment to be tested, and of any potential interactions of the study treatment(s) with already approved medical treatments.

Cleaning

Removal of contamination or soils from an item or surface to the extent necessary for its further processing and its intended subsequent use.

CMDCAS

Canadian Medical Devices Conformity Assessment System.

CMDR

Canadian Medical Device Regulation.

Conformity

Fulfilment of a requirement or meeting a requirement.

Conformity Assessment Body (CAB)

A body, other than a regulatory (competent) authority, engaged in determining whether the relevant requirements in technical regulations or standards are fulfilled.

CRO

A "contract research organisation", also commonly known as a "clinical research organisation", is a service organisation that provides support to the pharmaceutical and biotechnology industries. CROs offer clients a wide range of "outsourced" pharmaceutical research services to aid in the drug and medical device research and development process.

D

Data Integrity

Refers to the degree to which data is reliable and without error. Data must be accurate, attributable, contemporaneous, original, legible and available. A breach of data integrity occurs when any person manipulates or distorts data and submits the results of that data as valid.

Dead Leg

A dead leg in the world of piping terminology refers to an area of piping where there is insufficient flow or a tendency for water build-up or stagnation. The formal definition of a dead-leg states that pipelines for the transmission of purified water for manufacturing or final rinse should not have an unused portion greater in length than six diameters (6D rule) of the unused portion of pipe measured from the axis of the pipe in use.

Debugging

The process of locating, analysing, and correcting suspected faults or machine issues.

Design Controls

Design controls are a collection of practices and procedures that are incorporated into the design and development process for a product such as a medical device. They provide a structure and clear path from the user needs assessment to product delivery through a step-by-step process. Design controls ensure proper assessment of the design is completed during the design and development phase. Design controls are a requirement of quality systems such as 21 CFR Part 820 (medical devices), and for certain classes of devices and per ISO 13485 - Quality Management Systems.

Decommissioning

When a system is taken out of production service and stored in an adequate environment for potential future use.

Depyrogenation

A thermal process used to destroy or remove pyrogens (endotoxins). Typically, primary packaging components such as glass vials are subject to depyrogenation.

Detection Limit

The lowest amount of analyte in a sample that can be detected but not necessarily quantitated as an exact value for an individual analytical procedure. (Ref: ICH Q2.)

Design History File

The DHF is a repository for all of the documentation generated as a result of the design control process. The DHF serves as a complete record of the design.

Design Validation

Establishing by objective evidence that device or product specifications conform to user needs and intended use(s) defined in design documentation.

Debarment

The FDA has the authority to disqualify or remove researchers from conducting clinical testing of new drugs and devices when the agency determines that the researcher has repeatedly or deliberately not followed the rules intended to protect study subjects and ensure data integrity. Further, the FDA can disqualify a clinical investigator who has repeatedly or deliberately submitted false information to the agency or study sponsor in a required report.

Under its statutory debarment authority, the agency may also ban or "debar" from the drug industry individuals and companies convicted of certain felonies or misdemeanours related to drug products. Once individuals have been subjected to debarment, they may no longer work for anyone with an approved or pending drug product application at FDA. Debarred companies may no longer submit abbreviated drug applications.

Design Qualification (DQ)

The documented verification that the proposed design of the equipment is suitable for the intended purpose. DQs are typical deliverables for facilities, systems and equipment and/or processes.

Design Space

The multidimensional combination and interaction of input variables, e.g. material attributes and process parameters that have been demonstrated to provide assurance of quality. Working within the design space is not considered as a change.

Directives

Directives are legal requirements. These must be met by manufacturers. Standard such as ISO 13485 help companies meet the requirements of directives, such as "Guidelines Relating to the Application of the Council Directive 93/42/EEC on Medical Devices."

Direct Impact (System)

A system that is expected to have a direct impact on product quality. These systems are designed and commissioned in line with good engineering practice (gep) and, in addition, are subject to qualification and validation. Such systems include HVACs and clean utilities such as WFI (Water-for-Injection)

Diffusion Blending

A process in which particles are reoriented in relation to one another when they are placed in random motion and interparticular friction is reduced as a result of bed expansion (usually within a rotating container). Also referred to as tumble blending.

Deviations

A deviation can be simply described as an unintended event which causes a test or verification to fail to meet expected acceptance criteria.

Degree of Invasiveness

A device, which in whole or in part, penetrates inside the body either through a body orifice or through the skin surface, is invasive. Invasiveness is generally categorised as invasive of a body orifice (including the surface of the eye), surgically invasive devices and implantable devices.

Device Master Record (DMR)

A compilation of records containing the procedures and specification for a device. The contents of a DMR can contain local procedures such as SOPs and work instructions along with global or divisional specifications used to detail manufacturing processes, intermediate product or final product.

Drug Product

The dosage form in the final immediate packaging intended for marketing. The finished dosage form that contains a drug substance, generally, but not necessarily in association with other active or inactive ingredients. (FDA)

Duration of Contact

In determining the classification of a device, the duration that the device is in continuous contact with the patient is defined as transient, short term or long term. The longer the device is in contact with the patient or user, the greater the risk and therefore this has to be taken into account when determining classification. Continuous use is defined in MEDDEV 2.4/1 as the uninterrupted actual use for the intended purpose. Where use of a device is discontinued in order that the device is immediately replaced with an identical device (e.g. replacement of a urethral catheter) this shall be considered as continuous use of the device.

E

Electronic Signatures

Electronic signatures are computer-generated character strings that count as the legal equivalent of a handwritten signature. The regulations for the use of electronic signatures are set out in 21 CFR Part 11 of the FDA. Each electronic signature must be assigned uniquely to one person and must not be used by any other person. It must be possible to confirm to the authorities that an electronic signature represents the legal equivalent of a handwritten signature. Electronic signatures can be biometrically based or the system can be set up without biometric features.

Encapsulation

The division of material into a hard gelatine capsule. Encapsulators should all have the following operating principles in common: rectification (orientation of the hard gelatine capsules), separation of capsule caps from bodies, dosing of fill material/formulation, re-joining of caps and bodies, and ejection of filled capsules.

Endotoxin

A pyrogenic product (e.g., lipopolysaccharide) present in the bacterial cell wall. Endotoxin can lead to reactions in patients receiving injections ranging from severe fever to death.

Equipment Qualification

Qualification means the process to demonstrate the ability to fulfil specified requirements. EQ consists of proving and documenting that equipment or ancillary systems are properly installed (installation qualification, iq), work correctly (operations qualification oq), and the different sub-systems work together as a system (performance qualification pq) and actually lead to the expected results. Qualification is part of validation, but the individual qualification steps alone do not constitute a validated process.

Excipient

Substances other than the API which have been appropriately evaluated for safety and are intentionally included in a drug delivery system to provide a specific role in manufacturing, shelf-life or physical property.

Equipment Range

The full range that equipment is capable of performing, as per the manufacturer specification and tolerances. (a process may not utilise the full equipment range, operating over a narrower range).

F

Factory Acceptance Testing (FAT)

An FAT or Factory Acceptance Test is an engineering activity that inspects and verifies that the equipment or system meets the requirements of the URS.

Failure Mode and Effects Analysis (FMEA)

A risk assessment tool that provides for an evaluation of potential failure modes and their likely effect on outcomes and/or product or process performance in order to prioritise risks and monitor the effectiveness of risk control activities. It is often used to identify areas within a given process, product, or system that render it vulnerable.

FDA 483s

An FDA 483 letter typically includes a summary of findings and observations in relation to an audit or inspection where the FDA representatives have reason to believe GMP or other regulations have been violated or are not being met. In response to an FDA 483 letter, the company should address each item and provide a timeline for correction or request clarification of what changes are required.

Functional Design Specification (FDS)

A functional design specification is a document that specifies how particular requirements are met – this can be a combination of how the equipment/process operates mechanically/automatically etc. An FDS is typically written in response to a URS

Fluid

A fluid is a substance that undergoes continuous deformation when subjected to a shearing force.

G

GAMP

Good Automated Manufacturing Practice (GAMP) is a set of guidelines for manufacturers and users of automated systems in regulated industries, specifically the medical device, pharmaceutical and biopharmaceutical industries. The application of GAMP and validation of automated systems in manufacturing helps ensure that regulated medical devices and medicinal products have the required quality and are manufactured according to good practices, meet regulatory and legal requirements and ensure patient safety.

Good Documentation Practices, GDP

The handling of written or pictorial information describing, defining, specifying and/or reporting of certifying activities, requirements, procedures or results in such a way as to ensure data integrity.

Granulation

A process of creating granules. The powder morphology is modified through the use of either a liquid that causes particles to bind through capillary forces or dry compaction forces.

Grade A Areas

Aseptic processing areas, critical in nature where sterile products are exposed to the environment receiving no further sterilisation. High-risk operations (for example aseptic stopperage, filling, loading of the lyophiliser) occur in Grade A areas. They are considered ISO 5 under both dynamic and static conditions.

Grade B Areas

Aseptic processing areas where the sterile product is protected from the environment. Grade B processing areas are the background environments for Grade A areas and are considered ISO 7 environments in the dynamic state and ISO 5 environments under static conditions.

Grade C Areas

Non-critical areas where bulk product or materials are exposed to the environment, yet final sterilisation has not yet been performed. Grade C areas are support areas for non-sterile production activities; purification, formulation, and preparation of components, equipment, etc. for sterilisation. They are considered ISO 8 (Class 100,000) environments in the dynamic state and ISO 7 (Class 10,000) environments under static conditions.

Grade D Areas

Non-critical production areas, support areas, airlocks, or corridors. They are support areas for non-sterile production activities in closed systems; cell culture, or buffer and media preparation areas. Grade D airlocks are used for the movement of product, materials and personnel into classified areas.

GHTF

Global Harmonisation Task Force.

GxP

GxP is a general term for good practice with regard to quality guidelines and regulations. These guidelines are used in many fields, including the pharmaceutical, medical device and food industries. X is used as an umbrella letter representing different subjects or disciplines in industry. Some prime examples include GLP (Good Laboratory Practice), GDP (Good Documentation Practice), GEP (Good Engineering Practice) and GMP (Good Manufacturing Practices). Furthermore, the use of a lower case "c" as a prefix indicates "current" or "up-to-date".

H

Harm

Damage to health, including the damage that can occur from loss of product quality or availability.

High Level Risk Assessment (HLRA)

A high-level risk assessment that can be used at the beginning of a project to estimate the risk, such as the risks involved with bringing in new computerised/automated equipment.

HVAC

Heating, ventilation and air-conditioning (HVAC) systems are used to control the environmental conditions within an area or manufacturing facility. HVAC systems also provide comfortable conditions for operators based in the manufacturing environment. Temperature, relative humidity (RH) and ventilation should not adversely affect the quality of products during their manufacture and storage, or the proper functioning of equipment.

Hydrogel

A biomaterial made up of a network of polymer chains that are highly absorbent and as flexible as natural tissue.

I

ICH

International Conference on Harmonisation of Technical Requirements for Registration of Pharmaceuticals for Human Use.

Intended Purpose

Intended purpose means the use for which the device is intended according to the data supplied by the manufacturer on the labelling, in the instructions and/or in promotional materials. (Chapter I section 1 of Annex IX of Directive 93/42/EEC.)

Impurity

Any component of the new active pharmaceutical ingredient which is not the chemical entity defined as the new active pharmaceutical ingredient **_or_** any component present in the active pharmaceutical ingredient or final product which is not the desired product, a product-related substance, or excipient including buffer components.

Invasive Device

A device, which, in whole or in part, penetrates inside the body, either through a body orifice or through the surface of the body.

IQ/OQ

Equipment IQ/OQ is defined as establishing documented evidence that all key aspects of the process equipment installation adhere to the manufacturer's approved specifications and any recommendations of the supplier of the equipment are suitably considered. The process/equipment must also operate as intended and all user requirements must be adequately fulfilled.

IFU

Instructions for Use.

Injunction (Plant)

An injunction is a judicial process initiated to stop or prevent violation of the law, such as to halt the flow of violative products in interstate commerce and to correct the conditions that caused the violation to occur. (FDA 21 U.S.C. 332; Rule 65, Rules of Civil Procedure.)

If a firm has a history of violations and has promised correction in the past but has not made the corrections, the injunction is more likely to succeed. However, the freshness of the evidence is critical.

For an injunction action to be credible in the eyes of the Department of Justice (DOJ), the U.S. Attorney and the court, the evidence must be current. Timeliness is an important factor when considering an injunction action, with or without a Motion for Preliminary Injunction or a temporary restraining order (TRO). However, case quality and credibility must not be sacrificed to meet guideline time frames. The purpose of the guideline time frames is to limit, as much as can reasonably be expected, the need to update evidence. Updating entails extra work at all levels of the case development and review process and more importantly, delays obtaining an injunction which is intended to stop violations that adversely affect the safety or quality of products in commerce.

ISO

International Organisation for Standardisation. Agency responsible for developing international standards, e.g. ISO 13485 Medical Devices.

Isolator

A sealed enclosure, which provides full physical separation between the critical processing zone and the other surrounding processing zones. The internal surfaces of the isolator and its contents are decontaminated in accordance with defined objectives, by highly effective cycles, e.g. vaporised hydrogen peroxide. The enclosure must be capable of preventing ingress of contaminants by means of physical interior/exterior separation, and be capable of being subject to reproducible interior bio-decontamination.

Isoelectric Precipitation

Isoelectric precipitation works by reducing the electrostatic forces to near zero, allowing the proteins to precipitate out.

ISO 13485

ISO 13485 is an ISO standard, published in 2003, that represents the requirements for a comprehensive management system for the design and manufacture of medical devices.

ISO 14971

An ISO standard, published in 2007, that provides a framework and requirements for a risk management system for medical devices. This standard establishes the requirements for risk management to determine the safety of a medical device by the manufacturer during the product life cycle.

ISO 9001

ISO 9001 is an ISO standard that represents the requirements for quality management systems. It is used across industries and is not specific to medical devices like ISO 13485.

Item Master

The item master is a record of all components that a manufacturer buys, builds or assembles into its products. The item master includes information like the size, shape, material, manufacturer, manufacturer part number and vendor for each component.

IVD

In vitro diagnostic tests are medical devices intended to perform diagnoses from assays in a test tube, or more generally in a controlled environment outside a living organism.

IVDD

The in vitro diagnostic device directive delineates requirements that in vitro diagnostic devices must meet before they can be sold in the EU market.

Intermediate

A material produced during steps of the processing of an API that undergoes further molecular change(s) or purification before it becomes an API.

J

JIT (Just in Time)

A strategy used to monitor inventory levels with the goal of reducing inventory and associated carrying costs.

K

Kanban

A scheduling system that advises manufacturers what to produce, when to produce and how much to produce. Pioneered by Toyota, the approach is based on demand. Inventory is replenished only when visual cues like an empty bin, trolley or cart show that it's needed.

L

Laminar Flow

Laminar flow is when fluid particles move in parallel layers at a constant velocity.

Life Cycle (Validation)

The validation life cycle refers to the requirement to control and document all validation activities from conception and URS stage to the retirement of equipment or a process. The life cycle approach ensures compliance throughout the life of the process/equipment while maintaining a validated state throughout the application of change control.

Linearity

The ability of an analytical procedure (within a given range) to obtain test results that are directly proportional to the concentration (amount) of analyte in the sample.

Line Clearance

The act of performing and documenting the removal of materials from a production or packaging line and cleaning prior to the introduction of a new batch or lot.

Lyophilisation (Freeze Drying)

Lyophilisation is the removal of ice or other frozen solvents from a material through the process of sublimation and the removal of bound water molecules through the process of desorption.

M

Maximum Allowable Carry Over (MACO)

The amount of allowed product residue (carry-over) from lot-to-lot, batch-to-batch. This limit is based on the most conservative or lowest level of three MACO calculation methods: (1) limited based on toxicity, (2) limit based on smallest therapeutic dose, and (3) worst-case dose.

Measurement Capability Index (MCI)

The Measurement Capability Index (MCI) represents the capability of the measurement system. It is used to evaluate the capability of the gauge to classify product against predetermined specifications.

Measurement System Analysis (MSA)

A study to determine the degree of error involved in measuring the given parameter. The measurement system involves the combination of operations, procedures, gauges, instruments, environmental conditions, people and software.

Medical Device

A medical device is "an instrument, apparatus, implement, machine, contrivance, implant, in vitro reagent, or other similar or related article, including a component part, or accessory which is:

- recognised in the official National Formulary, or the United States Pharmacopeia, or any supplement to them,

- intended for use in the diagnosis of disease or other conditions, or in the cure, mitigation, treatment, or prevention of disease, in man or other animals, or

- intended to affect the structure or any function of the body of man or other animals, and which does not achieve any of its primary intended purposes through chemical action within or on the body of man or other animals and which is not dependent upon being metabolised for the achievement of any of its primary intended purposes."

Medicinal Drug Products (Finished Products)

Finished dosage forms (e.g. tablet, capsule, or solution) that contain the active pharmaceutical ingredient usually combined with inactive ingredients. Medicinal products are intended to furnish pharmacological activity or other direct effect in the diagnosis, cure, mitigation, treatment, or prevention of disease or to affect the structure and function of the body.

MDD

The Medical Device Directive is intended to harmonise the laws relating to medical devices within the European Union. Medical Device Directive 93/42/EEC was most recently reviewed and amended by 2007/47/EC.

MHRA

The Medicines and Healthcare Products Regulatory Agency (MHRA) is the UK government agency which is responsible for ensuring that medicines and medical devices work and are acceptably safe.

MSDS

Material Safety Data Sheet.

N

NCR

Non-Conformance Report.

NIH

National Institutes of Health (U.S.)

NOEL

No Observed Effect Level. In relation to cleaning validation.

Non-Conformity

A deficiency in a characteristic, product specification, CQA, process parameter, record, or procedure that renders the quality of a product unacceptable, indeterminate, or not according to specified requirements.

Non Parametric Data

Where the type of data is non-variable. Also referred to as attribute data, e.g. visual inspection resulting in a PASS/FAIL result.

Notified Bodies

A notified body is a certification organisation which the national authority (the competent authority) of a member state designates to carry out one or more of the conformity assessment procedures or audits described in the annexes of the medical devices directives or GMP legislation.

NPI (New Product Introduction)

The market launch or commercialisation of a new product. NPI takes place at the end of a successful product development project.

O

Open System

An environment in which system access is not controlled by persons who are responsible for the content of electronic records on the system (21 CFR, Part 11).

Outlier

A test result that is statistically different compared to a set of other test results obtained from the same sample or samples from the same lot of material.

Out-of-Specification

A recorded result that falls outside the established specification(s) or acceptance criteria.

Out-of-Trend

Analytical result, which is within specification or acceptance criteria, but different from those usually obtained or expected. Out-of-trend results should be investigated by the same general principles as out-of-specification results.

Quantitation Limit

The lowest amount of analyte in a sample which can be quantitatively determined with suitable precision and accuracy for an analytical procedure. The quantitation limit is a parameter of quantitative assays for low levels of compounds in sample matrices and is used particularly for the determination of impurities and degradation in products.

Overall Equipment Effectiveness (OEE)

A calculation for measuring the efficiency and effectiveness of a process by equipment breaking it down into three constituent components. (The OEE factors: Availability x Performance x Quality.)

Overkill

A sterilisation process that is demonstrated as delivering at least a 12 Spore Log Reduction (SLR) to a biological indicator having a resistance equal to or greater than the bioburden level.

P

Pan Coating

The uniform deposition of coating material onto the surface of a solid dosage form while being translated via a rotating vessel.

Particle Count Test

This test covers verification of cleanliness. Dust particle counts are measured. The number of readings and positions of tests should be defined in accordance with ISO 14644-1 Annex B5.

Performance Indicators

Measurable values used to quantify quality objectives to reflect the performance of an organisation, process or system, also known as performance metrics in some regions. (ICH Q10.)

Performance Qualification (PQ)

Establishing by documented evidence that the process, under anticipated (controlled) conditions, consistently produces a product which meets predetermined requirements.

Precision

The degree of agreement (scatter) between a series of measurements when a method is applied repeatedly to multiple samplings of a homogeneous sample or artificially prepared sample under the prescribed conditions. There are three types of precision; repeatability, intermediate precision and reproducibility.

Pressure Cascade

A process whereby air flows from one area, which is maintained at a higher pressure, to another area at a lower pressure.

Piping and Instrument Diagrams (P&IDs)

Engineering technical drawings that provide details of the connections and integration of equipment, services, material flows, plant controls and alarms. The P&IDs also provide the reference for each tag or label used for identification.

PMA

Premarket approval by FDA is the required process of scientific review to guarantee safety and effectiveness for Class III devices.

PMDA

The Pharmaceutical and Medical Devices Agency in Japan reviews applications for marketing approval of pharmaceuticals and medical devices. It also monitors their post-marketing safety and provides relief compensation for people who have suffered from adverse drug reactions from pharmaceuticals or infections from biological products.

PMS

Post marketing surveillance is the practice of monitoring a pharmaceutical drug or device after it has been released on the market.

Process Design

Defining the commercial manufacturing process based on knowledge gained through development and scale-up activities.

Process Qualification

Confirming that the manufacturing process as designed is capable of reproducible commercial manufacturing.

Process Window

The selected operating range of machine settings/parameters that will produce product to meet all quality and product specifications.

Product Recovery

Product recovery is a critical and important step in the process. It is also referred to as "downstream processing". It is often the most expensive step in the process. For recombinant-DNA derived products, purification can often account for 90% of the total production costs.

Prospective Validation

Prospective validation is when validation is done in advance of commercial manufacturing.

Procedures

Also known as Standard Operating Procedures (or SOPs), procedures give directions for performing certain operations.

Protocols

Protocols give instructions for performing and recording certain discreet operations. (Examples include engineering protocols, validation protocols etc.)

Pure

A term typically used within pharmaceutical manufacturing, a product or substance is pure if it is free of contaminants, foreign matter, chemicals and harmful microbes.

Q

Quality Management System

A Quality Management System, often abbreviated to QMS, is any system based on a collection of business processes that are primarily focused on providing safe and quality products that consistently meet customer requirements.

Quality

The degree to which a set of inherent properties of a product, system, or process fulfils requirements. (ICH Q9.)

Quality by Design

This is a systematic approach that begins with predefined objectives and emphasises product and process understanding and process control, based on sound science and engineering principles.

Quarantine

The status of materials isolated physically or by other effective means pending a decision on their subsequent approval or rejection.

Quality Policy

A document in which a company or organisation outlines their commitment and approach to quality. It usually sets out how they plan to achieve a high and consistent standard of quality. It should in some way speak to the customer or end user.

Qualification Plan

A Qualification Plan (QP) describes all the qualification measures and at which stage of the qualification the verification will be completed. It typically contains detailed descriptions of the necessary test measures and a description of the interdependencies of the individual tests. In some instances, there may not be a need or a requirement for a qualification plan. A validation plan can also serve to detail the qualification strategy.

QP

Companies that intend to manufacture or import medicinal products or intermediate products for use in clinical trials or for the EU market must appoint a qualified person in order to comply with EU good manufacturing practice standards.

QPM

Quality Policy Manual.

QSP

Quality System Procedure.

QSR

Quality System Regulations.

R

Range

Range is defined as the interval between the upper and lower measurements required. The minimum specified range should be within the equipment range and validated to operate at all points within the range.

Recall

As defined at 21 CFR 7.3(g), "recall means a firm's removal or correction of a marketed product that the Food and Drug Administration considers to be in violation of the laws it administers and against which the agency would initiate legal action, 2 21 CFR 806.2(h). e.g., seizure. Recall does not include a market withdrawal or a stock recovery." Recall does not include routine servicing. Recall also does not include an enhancement, as defined by this guidance.

Relative Humidity

The ratio of the actual water vapour pressure of the air to the saturated water vapour pressure of the air at the same temperature expressed as a percentage. More simply put, it is the ratio of the mass of moisture in the air, relative to the mass at 100% moisture saturation, at a given temperature.

Reusable Medical Device

A device intended for repeated use either on the same or different patients, with appropriate decontamination and other reprocessing prior to re-use.

Reusable Surgical Instrument

An instrument intended for surgical use by cutting, drilling, sawing, scratching, scraping, clamping, retracting, clipping or similar surgical procedures, without connection to any active medical device and which is intended by the manufacturer to be reused after appropriate procedures for cleaning and/or sterilisation have been carried out.

Re-Qualification

Requalification is designed to verify and ensure that the equipment/instrument/system is maintained in a qualified state after modification or after a stipulated time period (downtime).

Residual Risk

The risk level remaining after applying the identified controls on a high risk of harms and hazards manifestation.

Resolution

The smallest change in quantity that can be detected or provided by an instrument.

Residual Solvent

Organic volatile chemicals used or produced during the manufacture of APIs or excipients, or in the preparation of medicinal products.

Retain Samples

Samples that are kept for potential investigations and retests. It should be noted that retained samples are not a regulatory requirement as per Annex 10 or 21 CFR part 11.

Retrospective Validation

Retrospective validation is used for facilities or processes that have not completed formal validation. Historical data or a retrospective review can provide the evidence that the process or facility is operated as intended.

Rinse Sampling

Using a solvent to contact all surfaces of the sampled item to quantitatively remove target residue. The solvent can be water, water with pH adjusted, or organic solvent.

Right First Time

Right first time strives to create a culture of excellence. People are challenged with performing their tasks always in the correct manner to achieve the correct results always — right the first time.

Risk

The combination of the probability of occurrence of harm and the severity of that harm.

Risk Management

Risk management involves the systematic application of management policies, practices and procedures that identify, analyse, control and monitor risk. It is important to recognise that risk management should begin at the outset of the design and development phase of a project. The first step is to identify the user needs and intended use and application of the device.

RoHS

"Restriction of Hazardous Substances in Electrical and Electronic Equipment 2002/95/EC". An initiative that was adopted by the European Union (EU) in February 2003 and put into effect July 1, 2006.

Ruggedness

An indication of how resistant a test method or process is to typical variations in operation, such as those to be expected when using different analysts, different instruments and different reagent batches.

S

Scaffold

A structure of artificial or natural materials on which tissue is grown to mimic a biological process outside the body.

SKU

SKU is an acronym standing for **s**tock **k**eeping **u**nit. It represents a unique sales stock identifier.

Specifications

An approved document detailing the requirements with which the products or materials used or obtained during manufacture have to conform to. They serve as a basis for quality evaluation.

Specificity

The ability to assess unequivocally the analyte in the presence of components which may be expected to be present.

Stability

Stability studies are used to demonstrate and justify assigned expiration or retest dates.

5S

5S is a Japanese methodology of organising and storing items in a work or lab environment. It has been adopted by many Western companies as a tool to help maintain standards and reduce errors and mix-ups. The "5s" represents each stage of the method:

Sort

Sorting out any items that are not in use and removing them to a more appropriate area such as a storage facilities or the bin.

Set-in-Order

The idea behind "set-in-order" is to be always organised. It requires "a place for everything and everything in its place". By setting things in order, we can help to make live processing and testing more efficient and reduce the risk of errors, omissions and accidents.

Shine

Regular cleaning is an important practice and it is always helpful to "clean as you go."

Standardise

Implement standard practices through SOPs and training. Standardisation can also be applied to workstation layout.

Sustain

Make it a habit! After implementing a 5s methodology, it is only effective if continuous efforts are made to "sustain" the changes.

Sterility Assurance (SAL)

SAL or "sterility assurance level" refers to the probability of a single viable microorganism occurring on an item after sterilisation. For a terminally sterilised medical device to be designated as "sterile", the minimum sterility assurance level must be SAL = 10^{-6} or better. When applying this quantitative value to assurance of sterility, an SAL of 10^{-7} has a lower value but provides a greater assurance of sterility than an SAL of 10^{-6}.

T

Tableting

The reconstitution of a powder blend in which compression force is applied to form a single unit dose (tablet).

Tableting Press

Tablet press subclasses primarily are distinguished from one another by the method that the powder blend is delivered to the die cavity. Tablet presses can deliver powders without mechanical assistance (gravity), with mechanical assistance (automation), by rotational forces (centrifugal), and in two different locations where a tablet core is formed and subsequently an outer layer of coating material is applied (compression coating).

Traceability Matrix

A traceability matrix is a document that links the user requirements and specifications to where the verification and testing have been documented within the validation activities. It also illustrates that all user requirements are traceable to the evidence-based test.

Turbulent Flow

Turbulent flow is when the movement of fluid particles are varying in velocity and direction.

U

Uniform

The product is manufactured consistently and will have the same quality between batches manufactured on different days.

UDI, Unique Device Identification

The UDI is a series of numeric or alphanumeric characters that is created through a globally accepted device identification and coding standard. It allows the unambiguous identification of a specific medical device on the market.

Uninterrupted Power Supply

An uninterruptible power supply (UPS) is a system for buffering the main power supply. If the power supply fails, the battery of the UPS supplies the required power. When the power supply returns, the UPS battery stops supplying power and is recharged.

Unit Operation

Unit operations are the individual steps in the process that modify materials and their properties at each step of the process. Each unit operation comes together to create a complete process.

User Requirement Specification, URS

The URS is a critical document that defines the requirements of a particular system, equipment or process. Requirements such as the functional and operational aspects of the system are typically documented here.

USP

United States Pharmacopeia.

V

Validation

Validation is confirmation via documented evidence that the particular requirements for a specific intended use can be consistently fulfilled under anticipated conditions.

Validation Master Plan

A document providing information on a company's validation work programme. It typically details timescales for the validation work to be performed along with the key deliverables.

Verification

Verification means confirmation by examination and provision of objective evidence (i.e. documentation) that the specified requirements have been fulfilled.

Vaporised Hydrogen Peroxide (VHP)

Vaporisation of liquid hydrogen peroxide which results in a mixture of VHP and water vapour. The VHP mixture is used to decontaminate isolators.

W

Warning Letter

A warning letter is a correspondence that notifies regulated industry about violations that FDA has documented during its inspections or investigations.

WEEE Directive

Waste Electrical and Electronic Equipment Directive. European Community directive 2002/96/EC where manufacturers are responsible for disposing of electrical/electronic waste.

WFI (Water-for-Injection)

WFI is sterile and pyrogen-free water containing no less than 10 CFU/100ml (Colony Forming Units) with a sample size of between 100 and 300 ml and an endotoxin level < 0.25 EU/ml.

WHO

World Health Organisation.

WI

Work Instructions.

Witnessed By

When signed or initialled is legal proof that the individual signing is physically present and observes the step, calculation, or operation being performed by someone else, and that all entries of data are true and accurate.

Worst Case

A set of conditions or parameters which, in combination with product specification or attributes at their limits, pose the greatest challenges to the process.

X
--
Y
--
Z

Zone Classification

Zone classification refers to GMP areas which include controlled (aka classified) and non-controlled manufacturing areas. Areas may be classified based on EU Grades A–D and/or ISO Class 5–8 (in the US - Class 100–Class 100,000 areas).

The Biotechnology HANDBOOK for Engineers

www.ingramcontent.com/pod-product-compliance
Lightning Source LLC
Chambersburg PA
CBHW081113180526
45170CB00008B/2821